The **Drinking Water** Book

The **Drinking Water** Book

How to Eliminate Harmful Toxins from Your Water

Second Edition

COLIN INGRAM

CELESTIAL ARTS
Berkeley | Toronto

Celestial Arts
an imprint of Ten Speed Press
PO Box 7123
Berkeley, California 94707
www.tenspeed.com

Distributed in Australia by Simon and Schuster Australia, in Canada by Ten Speed Press Canada, in New Zealand by Southern Publishers Group, in South Africa by Real Books, and in the United Kingdom and Europe by Publishers Group UK.

Cover design by Catherine Jacobes
Text design by Katy Brown

Library of Congress Cataloging-in-Publication Data
Ingram, Colin, 1936–
    The drinking water book : How to Eliminate Harmful
    Toxins from Your Water / Colin Ingram. — 2nd ed.
        p. cm.
        Includes index.
    ISBN-13: 978-1-58761-257-2
    ISBN-10: 1-58761-257-7
    1. Drinking water—Contamination—United
    States. 2. Water quality—United States.
    3. Drinking water—Purification. 4. Drinking
    water—Health aspects. I. Title.

    RA592.A1I54 2006
    363.6'1—dc22   2006011955
Printed in the United States of America
First printing, 2006
1 2 3 4 5 6 7 8 9 10 — 10 09 08 07 06

CONTENTS

Quick Tips

Are you the type of reader who wants some information quickly?
Okay, try this:

- If you get your drinking water from a tap, let the water run at full flow for ten seconds, then slow it down to half flow or less to fill your container. Running at full flow will flush out pollutants that have attached to (or grown on) the faucet components, and reducing the flow will make it less likely that any other pollutants will be detached as you fill your container.
- If you want water for a hot drink, draw cold water from your tap and heat it. Resist the temptation to draw hot water from the tap, as there's a greater chance of pollutants coming from your hot water heater.
- If you use any kind of water filter that is portable or detachable, store the filter in your refrigerator when it's not in use. This will greatly slow the growth of microorganisms in the filter.
- Once bottled water has been open for two days, store it in the refrigerator.
- City, town, and rural water utilities are required to send

all customers an annual water quality report. If you get your water from a public water supply and you have not received a report, ask them for one—it's free. The report should alert you to any contaminants detected above federal or state permitted levels.

- If your water is from a private well, the most common health risk is from bacterial contamination. Have your water tested for the presence of bacteria at least once a year.
- Fluoridation: There are some sound arguments for and against. On balance, I recommend against it.
- Almost all water utilities disinfect water with chlorine (sometimes together with ammonia). This does a good job of killing almost all microorganisms in the water, but the chlorine itself poses a certain degree of health risk. Don't drink chlorinated tap water on a regular basis.
- Is distilled water good or bad for you? The water itself is okay, but be careful of the containers you store it in.
- Is bottled water really safer than tap water from a public water supply? It depends on the type and brand of bottled water—some are and some aren't.

All of the above and hundreds more details are covered in this book. Hopefully, these quick tips have *wetted* your appetite.

Does Anyone Know What They're Talking About?

Some years ago I was living in a mountainous rural area far removed from all obvious sources of pollution. Yet in this pristine region there was an epidemic of cancer among the scattered residents. After much publicity and badgering, the state was forced into investigating the problem. The result was a "gray wash." That's when the investigating body admits there may actually be a problem but announces the results of the investigation over an extended period of time in order to dilute citizen response, and they suggest that there are many causes of the problem so that it's hard to take action against any one of them.

But the kinds of cancer occurring in the region were known to be environmentally induced, so something in the local area was contributing to the epidemic. Finally, traces of a chemical were found in our supposedly pure water supplies—a chemical declared to be harmless by the manufacturer and all of the government involved. This led me on a quest for information about whether the chemical was really harmful and if so, what to do about it.

The journey took me to public waterworks, county and state health agencies, testing laboratories, universities, and the Environmental Protection Agency (EPA). It expanded to include toxicologists and epidemiologists, bottled water companies, and manufacturers of water purifiers. My quest culminated in a multiyear research program (with no grants or other assistance to prejudice the results) and the founding of a drinking water research center. This book is one result of that research.

As a part of that research I discovered many things that the general public is unaware of. Here are just a few of them:

- Although most public health officials claim that your drinking water is safe, they're only guessing. Their guesses are as informed as they can be in many cases, but no one actually knows what "safe" is, and "safe" levels of pollutants are often based on incomplete data.
- Federal and state standards for drinking water safety are inadequate in several ways: They do not cover all of the toxic substances that may be in your water; many smaller public water systems are exempt; and almost no studies have been done on the increased toxicity caused by combinations of pollutants in water. In many cases, standards have been set at levels that accommodate industry rather than protect public health.
- Right now, there are trace amounts of chemicals known to cause cancer in essentially every public water supply in the country.
- Some water purifiers for home use create and add new pollutants to the water even as they remove other pollutants.
- Performance claims by some water purifier manufacturers are biased, misleading, or irrelevant.
- Most of the water tests performed for consumers in the United States test for aesthetic qualities (taste, smell, color, clarity) rather than for potentially dangerous pollutants.
- Popular magazines and consumer publications that test and/or report on water quality and water purifiers generally don't have staff with sufficient knowledge to provide consumers with useful and/or accurate information.

This is not to say that there aren't experts in specialized areas of water quality—there are. But try asking your local health department officials about the relative merits of different brands of distillers. Or ask a water filter salesperson for data on bacterial growth within filters. Or ask a bottled water dealer about studies that have shown that the plastic from some water bottles can migrate into the water and affect your immune system. Mostly, you'll get blank looks.

My purpose isn't to criticize these people who, by and large, are sincere and trying to be helpful, but to point out that the consumer has had nowhere to go for all the information needed to make intelligent decisions about safe drinking water. *The Drinking Water Book* puts it all together. It helps you to find out if you really need to do something about your drinking water—and what might happen if you don't. It describes the most cost-efficient ways to get better water, including some things you can do without spending a penny. If you decide to buy a water purifier, this book will help you understand all of your options, how to narrow your choices, and the best values available.

The Drinking Water Book includes tips and information you won't find in any other popular source, such as a summary of what is known about what kinds of water have been scientifically shown to help maintain and improve your health. While the information in this book is presented in a simple, easy-to-absorb form, it is based on years of extensive research on water quality and on actual testing of products, not just in laboratories but in actual installations in homes.

For many years, American consumers have been faced with frightening headlines about unsafe drinking water but haven't had any real guidelines on what to do about it. This book fills that gap.

In past years, information on drinking water and water treatment options has been hard to come by. Now, there is a surfeit of information on the Internet. Because so much information is available on the Web and because the majority of readers have access to it, this book is designed to be supplemented by the Internet.

Meanwhile, for more quick information, please go on to chapter 1.

CHAPTER 1

Questions and Answers

Why Should I Be Concerned about My Drinking Water?

Q: News stories about toxic substances in drinking water occasionally appear, but except in unusual cases, public health officials claim that there is no danger. Is there really a health threat from drinking tap water from a water utility?

A: It depends on where you live, the source of the water, and how the utility treats it. In some areas, toxic substances in tap water pose a serious health threat; in other areas, the tap water is relatively free of harmful substances. Also, water quality can vary from season to season and even from day to day. And while the Environmental Protection Agency (EPA) has established standards for more than one hundred pollutants, there are estimated to be in excess of two thousand different pollutants in our water supplies.

Q: How can there be polluted water in areas with no history of toxic substance use?

A: (1) There is no way to know for sure whether toxic substances have been used (or dumped) in a given area. (2) The water may originate from another area where toxic substances have been used. (3) Some of the chemicals used to treat water and make it safe are themselves harmful. (4) Plumbing within and outside of the house can also deposit pollutants into the water.

Who Is Responsible for Safe Drinking Water?

Q: Who is responsible for ensuring that our tap water is safe?

A: A combination of federal, state, and local government agencies, along with the water utility company that delivers your tap water.

Q: If tap water is unsafe for drinking, why does the government allow it?

A: Because it would cost a lot of money to make it safer; because many elected and appointed officials support industries that pollute water; and because many water utility companies resist their responsibility to make tap water safer.

Q: Isn't the purpose of water utilities to provide safe water?

A: Yes, but many of them, particularly the small systems, are oriented toward eliminating immediate health threats from the water (such as harmful microorganisms), and it is expensive to remove trace amounts of pollutants that affect long-term health.

Q: I've heard that industry, agriculture, and the military are still polluting water. Why doesn't the government force them to stop discharging pollutants into water supplies?

A: It would cost a great deal of money for these sectors to stop discharging all pollutants and, rather than finding solutions to the problem, special interests pressure the government to allow the pollution to continue.

Q: So, in spite of the taxes I pay, it sounds like I have to be responsible for the safety of my drinking water.

A: The various levels of government are responsible for safe drinking water, and they have done much, but not enough, to improve it. Because of this, you are, by default, ultimately responsible for ensuring that you and your family have safe drinking water.

Should I Have My Water Tested?

Q: How can I find out what is actually in my tap water?

A: If your water is from a water utility company, you should be able to get a free copy of test reports that show whether or not certain toxic substances are in the water. While these reports are helpful, they don't include a large number of other toxic substances that may be in the water.

Q: If I want to get a more comprehensive test of my water, will my city or county health agency, or other health agency pay for it?

A: In general, no, unless someone in your family has become sick and your doctor suspects water is the cause.

Q: How do I get a comprehensive test of my water?

A: You will have to pay for testing by a private laboratory. While most test labs charge very high prices, there are several automated labs that offer low-cost, comprehensive testing.

Q: So should I have my tap water tested?

A: It is often possible to infer a great deal about your water from information already on hand. For more information on how to find out what's in your water, see chapter 3.

What Should I Do about My Drinking Water?

Q: What are my options?

A: The first thing to do is to stop drinking chlorinated tap water as soon as possible. Then you have three basic choices: buy bottled water, buy water from a vending machine or water store, or install a water purifier at home. Even if you decide to install a water purifier, start drinking bottled water or water from a vending machine until the purifier is in place.

Q: Which is the best-quality drinking water: bottled water, water from a vending machine, or water treated with a purifier?

A: It depends on the particular kind you choose. If you buy the right kind of bottled water, its quality will be relatively high; however, a few bottled waters are worse than tap water. Most water from vending machines is of high quality, but it is important to pick the right machines. Water purifiers vary from extremely good to very bad; some purifiers actually add toxic substances to the water. The effectiveness of a purifier often depends on correctly matching it to your particular water conditions. In general, the best water purifiers will give you the purest water.

Q: If I want to drink bottled water, which kind should I buy?

A: Buy a well-known, major brand from a store that sells a lot of it. Choose a label that says "drinking water." If that's not available, choose "purified water" or "distilled water" until you can find "drinking water." Do not buy "spring water," "natural spa water," "natural mountain water," "well water" or any water with a label indicating that the water is from a natural source unless you are prepared to carefully check out the bottler's test data.

Q: But don't natural springs have the purest water?

A: Not necessarily. Water from natural sources sometimes contains naturally occurring toxic pollutants.

Q: If I want to drink water from a vending machine, how do I choose a machine?

A: Choose a popular location where the machine gets a lot of business. Look for a label or seal on the machine that indicates it is inspected by the county or other local health agency. And make sure your containers are clean.

Q: What should I know before buying a water purifier?

A: That's a more complicated question. Don't buy or rent any water purifier until you have read the sections of this book on purifiers.

(For more information on bottled water and water vending machines, see chapter 6. For more information on water purifiers, see chapters 7 through 13.)

CHAPTER 2

Water Pollutants and the Risk to Health

If you want to have safe drinking water and also avoid the potentially harmful effects of inhaling water pollutants or absorbing them through your skin, a first step is to understand the kinds of pollutants that may be in your water. Let's begin by dividing all water pollutants into two broad categories: nuisance pollutants and health-threatening pollutants. Nuisance pollutants are those that cause discomfort or inconvenience. They can cause water to taste, smell, or look bad, and they can render soap and washing less effective. First we'll look at the health-threatening pollutants, which fall into the following categories:

- Pathogens
- Toxic minerals and metals
- Organic chemicals
- Radioactive substances
- Additives

● Pathogens

Pathogens are harmful microorganisms such as bacteria, viruses, and parasites. They can cause such diseases as typhoid, cholera, hepatitis, flu, and giardiasis. The most common bacteria are closely monitored in public water supplies, both because they can be dangerous and because their presence is easily detected. Tap water from a public supply is generally free of dangerous concentrations of bacteria because they are killed when chlorine is added. Private wells may be contaminated with bacteria if the well is near a septic system, open to the air, or subject to seepage from chemicals or animal fecal matter.

Viruses are much smaller than bacteria and harder to detect. Viruses are very common in water. Research has shown that a teaspoon of relatively unpolluted lake water contains over a billion viruses. Although disinfection with chlorine, the method used by most utilities, probably kills the majority of viruses in water, no one knows for sure how many remain potent. It is more difficult to test for the presence of viruses in water than it is to test for bacteria, and most water testing laboratories don't have the ability to do this. While most waterborne viruses appear to be harmless to humans, some do cause cold and flu epidemics, and hepatitis. Harmful viruses can be present in public water supplies even though a water treatment plant is operating properly. Private wells will invariably contain viruses, although most are apparently harmless.

The third type of pathogens commonly found in water is protozoan parasites. Of these, the two most common and troublesome are giardia and cryptosporidium. In water, these parasites occur in the form of hard-shelled cysts. Their hard covering partially protects them from the chlorine disinfection that kills other microorganisms. Both giardia and cryptosporidium cause mild to severe gastrointestinal symptoms in healthy people. In people with impaired

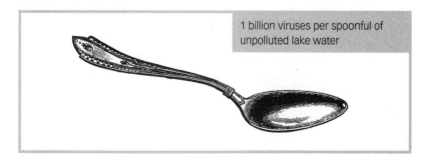
1 billion viruses per spoonful of unpolluted lake water

immune systems, they can be life threatening. Parasites can be present in public water supplies even though a water treatment plant is operating properly. Private wells that are deep and properly sealed from outside air will generally be free of parasites. Shallow wells can be subject to seepage from surface water and thus may contain parasites.

In general, while public water utilities do a good job of removing or killing most pathogens, some harmful organisms can remain in the water supply undetected. Deep, properly sealed private wells are generally free of pathogens, while shallow and/or unsealed wells are more prone to containing a variety of pathogens.

Here is a point to keep in mind that is rarely reported or discussed with regard to drinking water: live pathogens excrete toxic substances called *exotoxins*. But killing pathogens doesn't eliminate all potential toxicity. The disintegration of their bodies also produces toxic substances; these are called *endotoxins*. So in terms of potential harm to humans, there's a distinction between water that still contains the killed pathogens and thus their endotoxins, and water with the pathogens removed. Very few studies have been done on this, and the potential long-term, harmful effects of both endotoxins and exotoxins are largely unknown.

● Toxic Minerals

Toxic minerals are the harmful inorganic substances that are found in water supplies (*inorganic* means not derived from plant or animal matter). They include metals as well as common minerals in the form of rock, sand, and clay. It's important to distinguish between minerals that are harmful and those that are mere nuisances. The following minerals and materials are more of a nuisance:

- Calcium and magnesium, which cause hardness in water
- Iron and manganese, which cause staining
- Hydrogen sulfide gas, which causes a rotten egg odor

The minerals in water that are known to be harmful to health are:

- Aluminum
- Arsenic
- Asbestos
- Barium
- Cadmium
- Chromium
- Copper

- Fluoride
- Lead
- Mercury
- Nitrate
- Nitrite
- Selenium
- Silver

These toxic minerals and inorganic compounds occur naturally in water, and they also enter water from man-made sources. Some of them are more toxic than others. Cadmium, lead, and mercury have the greatest toxicity, and ingestion of even small amounts can be fatal. The ingestion of asbestos fibers is thought to increase the long-term risk of cancer. These fibers commonly occur in certain kinds of rock formations, especially those known as serpentine, a fairly common type of rock. Asbestos is also present in tap water wherever asbestos-cement water pipes are used to deliver water to customers. Unfortunately, that's just about everywhere in the United States. Even in very low-level doses, toxic minerals can over a period of many years cause disorders of the kidneys, bones, blood, and nervous system when ingested.

Nitrates and nitrites are mineral salts that can be harmful when present in high concentrations in water. This usually occurs in agricultural areas where large amounts of fertilizers are used or where livestock is raised, including grazing areas. High levels of nitrates and nitrites cause blue baby disease in infants and intestinal disorders in adults.

Toxic minerals can enter a water supply from naturally occurring sources in surface water or groundwater. They can also come from industrial discharges, runoff from urban and agricultural areas, and from the walls of water mains. They can even come from sources within the home. Metal pipes, joints, and plumbing fixtures, especially older sink faucets, are frequent sources of toxic mineral pollution.

In general, when a water source contains high levels of toxic minerals, water treatment plants do a good job of reducing levels to meet federal and state limits. However, there is a great deal of controversy as to whether or not these limits are sufficient to ensure safe drinking water.

♦ Organic Chemicals

Organic chemicals are substances that come directly from, or are manufactured from, plant or animal matter. Plastics, for example, are organic chemicals that are made from petroleum, which originally came from plant and animal matter. There are roughly one hundred thousand different manufactured, or synthetic, organic chemicals in commercial use today. They include synthetic fertilizers, pesticides, herbicides, paints, fuels, plastics, dyes, flavorings, pharmaceuticals, and preservatives, to name a few. Many of these chemicals are toxic, and thousands of them have been found in public water supplies. When synthetic chemicals are found in a water supply, the actual source of pollution may be a leaking gasoline tank or factory discharge many miles away; it may be agricultural runoff, herbicide that has been sprayed along highways, or any of hundreds of other legal or illegal sources. Often pollution of a water supply by synthetic chemicals has no obvious source—yet toxic chemicals are present nonetheless.

One category of organic chemicals is particularly dangerous. Volatile organic chemicals, or VOCs, are absorbed through your skin when you come in contact with water, as in a shower or bath. Further, hot water allows these chemicals to evaporate rapidly into the air, and they are harmful if inhaled. VOCs can be in any tap water, regardless of where you live or what your source of water is. If your tap water contains significant levels of these kinds of chemicals, there will be a health threat from the water even if you don't drink it.

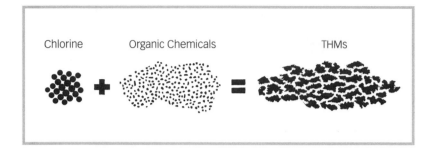

In addition to the organic chemicals that have gotten into water supplies, new and dangerous ones are created in the process of water treatment. Chlorine, which is added to essentially all U.S. public water supplies, combines with organic chemicals to form a category of toxic pollutants called trihalomethanes (THMs). THMs are known carcinogens—substances that increase the risk of getting cancer—and they are present at varying levels in all tap water from treatment plants.

◆ Radioactive Substances

Radioactive substances in water fall into two categories: radioactive minerals and radioactive gas. Radioactive minerals can be either naturally occurring or man-made. When naturally occurring, their source is typically an area where mining is going on or has gone on in the past. Uranium mining produces radioactive runoff, as you might expect. But other kinds of mines also enable radioactive minerals to enter water supplies. This is because mining exposes rock strata, most of which contain some amount of radioactive ore. Naturally occurring radioactive minerals can also enter water supplies through the operation of smelters and coal-fired electrical plants.

Man-made sources of radioactive minerals in water are nuclear power plants, nuclear weapons facilities, radioactive materials disposal sites, and docks for nuclear-powered ships. An unreported source of radioactive pollution comes from hospitals all over the country, which are allowed to dump low-level radioactive wastes into sewers. Some of those radioactive wastes eventually find their way into water supplies.

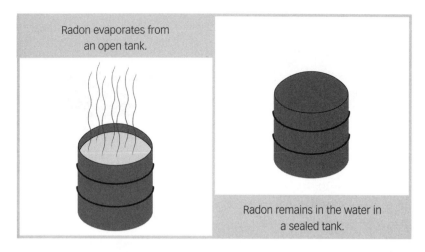

Radon evaporates from an open tank.

Radon remains in the water in a sealed tank.

If the source of your water is a reservoir, river, or mountain runoff that is near or downstream from any mining areas, nuclear facilities, or hospitals, you should pay special attention to finding out the level of radioactivity in your water.

While radioactive minerals in water may present a health hazard in these particular locations, a far more dangerous threat exists in the form of *radon*. Radon is a colorless, odorless, naturally occurring gas that is the byproduct of the decay of radioactive minerals. It is present in all water in minute amounts, and it is especially concentrated in water that has passed through rock strata of granite, shale, phosphates, or uranium ores.

Radon is a known carcinogen. When present in household water, it evaporates easily into the air and can be inhaled. The effects of radon inhalation are now believed to be a significant health threat.

Radon dissipates rapidly when water is exposed to air. Because of this, radon is not a threat from surface water—lakes, rivers, or aboveground reservoirs. Even when water comes from an underground source, radon is not a threat if the water is aerated (exposed to air) or processed through an open tank during treatment.

Radon may be a threat when groundwater comes into the home directly from an underground source—either from a private well or from a public water supply whose source is a well and whose treatment of the water does not include exposure to air. Because radon in water evaporates quickly into the air, the primary danger is from inhaling it from the air in an enclosed environment, not from drinking it. The Environmental Protection Agency (EPA) estimates that at least twenty thousand cases of cancer in the United States each year are caused by radon inhalation.

Radon gas enters a house in two ways: via the water system, from the soil beneath the home, and through cracks in the foundation and the like. Studies show that where there are high concentrations of radon in a building's air, most of the radon comes through the foundation and not from the water.

If radon is in tap water, all areas in a house where the water is exposed to turbulence and/or heat will have high concentrations of radon.

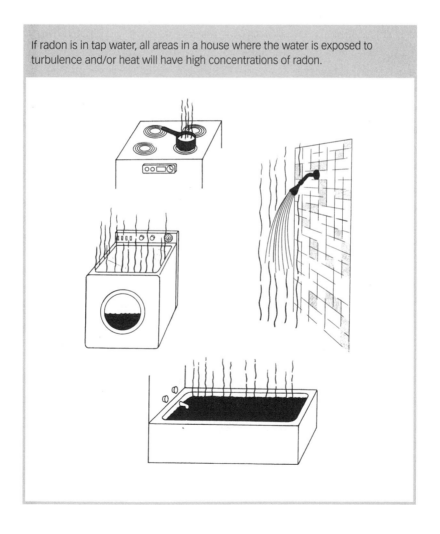

Radon concentrations within a house are highest in winter when all windows are shut. The problem is worsened by energy-efficient dwellings that have been tightly sealed to prevent heat loss. Basements have the highest radon concentrations. If radon is in the water, showers, baths, and cooking with hot water will cause high concentrations in the air.

Although radon is not, strictly speaking, a water problem, it is a very serious threat if you live in a region with high concentrations of radon in the groundwater.

Information on what to do about radon is in appendix A.

◆ Additives

All public water treatment plants, from small community systems to larger urban waterworks, add things to water. These are added for a variety of reasons, ranging from disinfection to enhancing effectiveness of treatment to improving the water's aesthetic qualities, and even (in theory) for public health benefits.

CHLORINE

The best-known additive is chlorine. Chlorine is used in almost every public water system in the United States, and it has been proven beyond any doubt to be an effective disinfectant. For decades it has been assumed that small amounts of chlorine in drinking water are safe. But convincing evidence, largely unreported to the public, has been accumulating for many years that shows a correlation between chlorinated water and a greater risk of serious diseases.

Ingesting small amounts of chlorine may itself pose a health threat, but in addition, overwhelming evidence indicates that the toxic chemicals formed when chlorine combines with organic chemicals in water *are* a serious health threat. These toxic chemicals are known as THMs and are present to some degree in all public water supplies, as I mentioned earlier in this chapter. Some public water systems have changed their primary disinfection agent from chlorine to chloramines, which are formed when chlorine is combined with ammonia. While chloramines form fewer THMs than chlorine by itself, they also form other compounds whose toxicity remains uncertain.

Recommendation: Do not drink chlorinated water on a regular basis.

FLUORIDE

Water fluoridation is another subject that has not been very well reported to the public. In the United States, the controversy about fluoridation has been going on since 1945, when fluoridated water was first introduced. In general, government agencies and the majority of the prestigious medical associations are strongly in favor of fluoridation. They claim that a small minority of dissenters have continued to raise false issues against fluoridation. Those against fluoridation, on the other hand, claim that vital research has been deliberately ignored and kept from the public.

Proponents cite studies proving that the addition of small amounts of fluoride to drinking water reduces tooth decay in children. The American Dental Association has endorsed this position for so long that it's taken as gospel truth. Dentists all over the United States give fluoride treatments; flouride is an ingredient in most toothpastes; schoolchildren in some districts are given fluorides rinses; and the majority of large cities fluoridate their drinking water.

Studies do show that rates of tooth decay in young people have gone down since fluoridation was introduced. On the other hand, rates of tooth decay have also decreased in cities (and countries) where the water isn't fluoridated.

An entire book would be required to cover this subject comprehensively, and even then the conclusions could be confusing. But here are some points for you to think about. For brevity's sake I'll call the proponents' views pro-F and the opponents' views anti-F.

> **Pro-F:** Countries adding fluorides to their water include the United States, Canada, Australia, Brazil, Chile, Columbia, Ireland, Israel, Malaysia, New Zealand, and the United Kingdom.

> **Anti-F:** These countries have rejected fluorides: Austria, Belgium, China, Denmark, Finland, France, Germany, Hungary,

India, Japan, the Netherlands, Norway, Sweden, and Switzerland. Europe in general is 98 percent unfluoridated.

Pro-F: Fluorides are naturally occurring compounds that are already present to some degree in all water supplies.

Anti-F: The naturally occurring fluorides in water supplies are not the same as the type of fluoride added to water. The type added to water is a waste product from the smokestack scrubbers of phosphate fertilizer plants, and it contains a number of other toxic wastes.

Pro-F: Fluoride added to water doesn't contain toxic wastes; fluorides are monitored for toxic elements and have to be in accordance with standards established by the American Waterworks Association and others.

Anti-F: The type of fluoride added to water supplies can be up to eighty-five times more toxic than naturally occurring fluoride.

Pro-F: The effects of naturally occurring fluoride and manufactured fluoride are the same, and over the past half century, thousands of studies have proved both the safety and the effectiveness of water fluoridation.

Anti-F: Over many years, "establishment" agencies and associations have systematically ignored or downplayed research that shows detrimental health effects from fluoridation.

Pro-F: The Centers for Disease Control has called water fluoridation one of the ten greatest health achievements of the twentieth century.

Anti-F: Many professionals at health-related government agencies disagree. For example, 1,500 professionals at the EPA made a thorough investigation into the pros and cons of fluoridation. Their report concluded that the health and welfare of the public is not served by the addition of fluoride to the public water supply.

Pro-F: The EPA has established the upper safety limit of the amount of fluoride in water to be 4,000 parts per billion (ppb) and has set the recommended level to be from 700 to 1,200 ppb. This ensures that that the amount of fluoride in the water supply is safe for drinking.

Anti-F: Fluorides have approximately the same toxicity as lead and arsenic. While the EPA has set the maximum contaminant level of lead at 15 ppb and arsenic at 50 ppb, it has set the fluoride level at 4,000 ppb.

Pro-F: Fluoridation is endorsed by the American Dental Association, the American Medical Association, the U.S. Public Health Service, the American Cancer Society, the Centers for Disease Control, the World Health Organization, and many others.

Anti-F: After studying the research, several prestigious groups have withdrawn their support, including the American Heart Association and the National Kidney Foundation.

Pro-F: In areas where the water supply is naturally high in fluorides, no additional fluoride is added to the water.

Anti-F: The problem is that there has been no attempt to determine overall fluoride intake. People ingest fluorides not only from drinking water but from canned and processed foods, juices and sodas, fluoridated toothpaste, fluoride rinses for children, and from fluoride residues in crops. No one knows how much of this poison we are ingesting.

Pro-F: There is such a preponderance of evidence for the benefits of fluoridation that the opposition can only be described as using junk science and having a political agenda.

Anti-F: There's now so much proof against fluoridation that it's amazing that proponents can keep saying that only a small minority oppose this public health measure. Plus, the government has no right to force a medical treatment (even a preventive one) on unwilling citizens.

Pro-F: The Supreme Court has ruled that providing a public service for the greater good of society is legal and justified.

So go the many arguments, and there are dozens more. If you want to delve deeper into the subject of water fluoridation, there's been a lot written on this topic. However, be warned; it's very hard to discern the truth on this issue. (After careful review of the research, I believe it's best to avoid drinking fluoridated water, if for no other reason than to reduce your overall toxic load, something I'll address later in this chapter.)

Recommendation: Do not drink fluoridated water on a regular basis.

FLOCCULANTS

In addition to chlorine, and sometimes fluoride, water treatment plants often add several other substances to water to improve the efficiency of the treatment. Flocculants are substances added to the water to make the particles in it clump together for more efficient removal by filtering.

Some of the most commonly used flocculants, called *polyelectrolytes*, have been banned in several other countries because some of their constituents are known to cause genetic mutations. The EPA classifies some of these flocculants as probable human carcinogens but still permits their use.

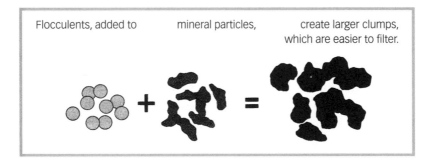

Flocculents, added to   mineral particles,   create larger clumps, which are easier to filter.

Taste, Smell, and Appearance

The taste, smell, and appearance of drinking water are normally mere nuisance factors. Brownish-colored water, for example, usually indicates the presence of harmless dirt or iron in the water, and a cloudy appearance may simply be no more than normal minerals in the water coming out of solution. But a bad taste or smell, or cloudy water, may also indicate the presence of harmful pollutants, such as bacteria. In general, if you notice an unusual taste or smell in your tap water, or a change in its appearance, contact your water utility company immediately and don't use the water for any purpose until you are given assurance that the water is safe.

Estimating Overall Health Risks

It would be impossible to estimate the health risks from the thousands of potential water pollutants individually, much less in combination. Is it possible to even come up with a generalized assessment of health risks?

In the United States, water treatment has concentrated on the elimination of acute (short-term) health risks, such as bacterial, viral, or parasitic infections, or acute poisoning from a sudden, large dose of lead or mercury. In this regard it has been fairly successful; serious acute waterborne diseases have been mostly eliminated. At the same time, those responsible for water quality have, until recent years, turned a blind eye to the long-term health effects of polluted water.

One of the excuses given for not confronting water pollution problems is that pollutants in very tiny amounts are often dismissed as inconsequential. Analogies are created for drama. For example, if the distance between San Francisco and Los Angeles is four hundred miles, then one billionth of that distance is less than ⅓₂ inch. But physical analogies are not particularly relevant to biological sensitivity. Just 1 part per *trillion* of hormones in human blood can cause changes. Similarly, 1 part per billion of a pollutant in water may do a lot of harm.

For certain individual pollutants, safety thresholds have been established, meaning that if a pollutant is below a certain concentration (say 1 part per million) in water, it's considered to have no harmful effect. Federal drinking water standards, known formally as the National Primary Drinking Water Regulations, establish limits to some ninety pollutants. These water standards apply to public water systems that provide water for human consumption with at least fifteen service connections or those that regularly serve at least twenty-five individuals. Public water systems include such entities as municipal water companies, homeowner associations, schools, businesses, campgrounds, and shopping malls. (For national drinking standards for specific pollutants and their legal safety thresholds, see the EPA link listed under EPA Drinking Water Standards in appendix C.)

The limit for each pollutant is referred to as the maximum contaminant level (MCL). With regard to the toxic THMs that are in every public water supply, these regulations currently establish 100 parts per billion as safe a level. This is an educated guess with a built-in safety factor. But, in fact, no one knows what level of THMs is safe over the long term—not government health officials, not scientists, not doctors. No one.

Here's another example: Inhaling asbestos is a known health risk, but not much is known about the health risks from ingesting asbestos in drinking water. Asbestos occurs in the form of microscopic fibers, and federal drinking water standards set an MCL of seven million asbestos fibers per liter of water. This MCL has been established not from any estimate of safety; rather, this is the best current laboratories can do in detecting asbestos in water. Further, even if there is no health risk from ingesting asbestos fibers at this concentration, health risks might arise from using water evaporators for cooling or humidity control, allowing waterborne asbestos fibers to enter the air.

So the safety of these thresholds is debatable, and they clearly don't take into consideration the complexities of how we use water. And even assuming the individual MCLs are safe, what if there are several dozen pollutants in your drinking water (which there probably are), with each of them at 1 part per million—is the water still safe? If, for example, your tap water contains toxic THMs, fluoride, pesticide residues, and asbestos fibers, how do these pollutants in combination, affect your long-term health?

Epidemiologists, toxicologists, and pathologists often talk about increased health risks from single sources of pollution, but they seldom address (at least publicly) overall environmental risk. Overall environmental risk is the effect on our health based on all the things we come into contact with each day. With regard to water quality, while there have been many studies of individual pollutants, there have been almost no studies of the combined, or synergistic, effect of pollutants. If, for example, your tap water contains toxic THMs, fluoride, pesticide residues, and asbestos fibers, how do these pollutants, in combination, affect your long-term health?

We find ourselves in a giant guessing game, and the stakes are human lives. In fact, there's no way to know for sure the health risk for any given individual from any single environmental cause or combination of environmental causes. Thus, we have to rely on common sense, and what common sense suggests is this: You can reduce your overall health risk by reducing your overall toxic load. In other words, reduce your exposure to all types of pollutants in your environment. Water quality is an important factor in that overall equation. Rather than accept so-called safe levels of pollution, you should try to remove or at least reduce levels of as many pollutants as possible from your drinking water.

CHAPTER 3

Finding Out
What's in Your Water

Before you can improve your tap water, you need to know what's
in it. This chapter describes what kinds of pollutants to look for
depending on your geographical location and the type of water
system that serves your home. It also discusses how to find out
more exactly, from your water company or through private testing, what actually is in your water. Be aware that to find out what's
in your water, you'll have to do some homework—there's no way
around it.

The Environmental Protection Agency (EPA) is authorized by the
Safe Drinking Water Act of 1974 to set quality standards for drinking water provided by public and private water utility companies. In
this chapter, one of the things we'll look at is whether or not your
tap water is in compliance with these standards.

◆ What's Likely to Be in Your Water in a City

City tap water is normally of a higher quality than tap water in
a small town or from a private water system. This is because city
water is provided by a water utility company with a customer base
large enough that it can afford sophisticated water treatment plants.
Although the water delivered by large urban water utilities occasion-

ally violates government standards, it is usually within these limits. When pollutant limits are exceeded, water utilities are obligated by law to inform the public. But sometimes they don't do this in a timely manner, or they may fail to do it altogether.

One way to find out for sure is to request a copy of your water utility's annual water quality report. This should be available without charge, and it will give you the yearlong average values for any pollutants in the water, plus the highest level of pollutants detected at any given time during that year.

Your next step is to find out, from your water utility, where your water comes from. Is it mostly or all from groundwater, or mostly or all from surface water? If it is from groundwater, be on the alert for high levels of radon and toxic metals. If your water company uses wells near any industrial facilities, gas stations, or landfills, be especially on the alert for organic chemicals in the water. The water utility's annual water quality report should detail any presence of these types of pollutants.

If your water utility's source is surface water (which includes lakes, rivers, reservoirs, and ponds), be on the alert for the presence of microorganisms, nitrates, and organic chemicals. Ask your city or county health department if there have been any local outbreaks of waterborne diseases, such as giardiasis.

Ask your water utility if they add *polyelectrolyte flocculants* to the water. This type of additive makes water filtration more effective, but the additives themselves frequently contain trace amounts of cancer-causing chemicals. While other industrialized countries ban or strictly regulate the use of these additives, the United States does not, so U.S. water utilities are free to use them.

Based on your inquiries, make a list of any pollutants that have been detected, even if they were detected at very low levels. If they were detected at all, chances are they'll be present again, sometimes at higher levels.

Here's a summary of the information you should get from your city water utility:

- A copy of their annual water quality report
- A list of any pollutants detected

- The locations and kinds of water sources
- Whether they add polyelectrolyte flocculants

Keep this information on hand to help guide you in deciding what to do about your water.

◉ What's Likely to Be in Your Water in a Small Town

Public water systems in small towns and rural areas generally don't have the sophisticated treatment plants that cities have. Also, small water systems have less frequent and less stringent testing requirements. In general, water systems in small towns are more often in violation of government standards than are large, urban systems, and there is a greater chance that pollutants are present in water from a small town's water system.

As a first step, ask your local water utility for its annual report of water quality. If this isn't available, ask to see test results for treated water from several different dates and in different seasons of the year—some pollutants are seasonal. These tests should clearly show if any pollutants exceed government standards and, if so, by how much.

The source of the water determines, to a large degree, what's likely to be in it. If your water supply comes from groundwater, it typically contains high levels of minerals, possibly including toxic minerals. As with all groundwater, the presence of radon is also a possibility unless the water is aerated as part of treatment before being delivered to customers. If you live in an agricultural area, be alert for the presence of nitrates, pesticides, herbicides, and other organic chemicals. On the positive side, groundwater is generally free of pathogens.

Water from a surface source generally has a low mineral content but is more prone to contamination from runoff and illegal dumping. If your water is from a surface source, inquire about any history of parasites in the water or local outbreaks of waterborne diseases. Surface water in an agricultural area is highly likely to be polluted with toxic chemicals such as nitrates, pesticides, herbicides, and fungicides. Also remember that the chlorine used as a

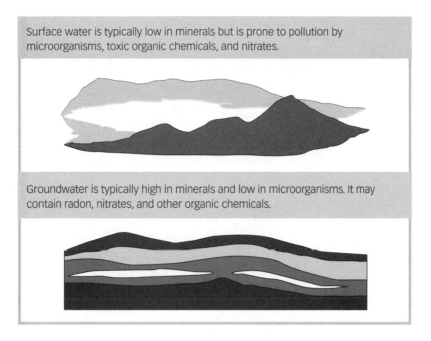

Surface water is typically low in minerals but is prone to pollution by microorganisms, toxic organic chemicals, and nitrates.

Groundwater is typically high in minerals and low in microorganisms. It may contain radon, nitrates, and other organic chemicals.

disinfectant in water treatment combines with organic chemicals to form toxic THMs. And because surface water in an agricultural area is likely to have relatively high concentrations of organic chemicals, the chances are greater of toxic THMs being present in high quantities.

Here's a summary of the information you should get from a small town water utility:

- A copy of their annual water quality report. If they don't provide one, make notes or entries from their logs of any pollutants detected
- The locations and kinds of water sources
- Whether they add polyelectrolyte flocculants

Keep this information on hand to help guide you in deciding what to do about your water.

◆ What's Likely to Be in Your Water in a Private System

In general, the explanations of risks from groundwater and surface water for city and small town public water systems apply to private systems. Probably the first and most important consideration with private systems is whether or not your system is sealed—that is, whether or not the water is exposed to outside air at any point from the well to your faucet. An unsealed well is susceptible to pollution from surface pathogens.

Many private water systems in rural areas are designed and constructed by property owners and don't have to meet health codes. Typical problem areas for pollution are unsealed wellheads, unsanitary springs, or unsealed or open holding tanks. The greatest health risk from these systems is the presence of algae, decaying plant matter, insects, and animal waste products, all of which lead to high levels of potentially harmful pathogens.

Another potential problem with private systems is the shallow well. In areas with high, year-round water tables, seepage from surface water can contaminate the well. This is typically more pronounced during rainy seasons.

Often, water from unsanitary systems looks and smells okay, and families may have been drinking this kind of polluted water without apparent harm. But high levels of bacteria and other biological pollution put stress on the human body and make it more susceptible to a variety of diseases seemingly unconnected to the water.

Since you can't ask for reports from a water utility if you have a private water system, you'll need to have the water tested yourself. You should have it tested annually for bacterial pollution, and you should also get a comprehensive multipollutant lab test once a year. The bacterial test should be done during the rainy season when there is the most surface water seepage. Stagger the times of the comprehensive tests so that water samples are taken at different times of the year. (Water testing is discussed in more detail later in this chapter.)

Because well water is typically high in mineral content there is a chance that it may contain toxic minerals as well as beneficial ones. Comprehensive lab tests should include testing for toxic minerals.

Another risk from a well is the presence of radon in the water. The risk of radon in well water varies considerably from region to region. If you live in a high-risk area, you should definitely test for radon. If you live in a low-risk area, radon may still be present. One way to find out if this is the case is to contact your county health department and ask if high radon levels have been found anywhere in your area.

◆ Pollutants from the Delivery System

Even if water treatment plants were totally effective in removing all pollutants from water sources, the water that reaches your home would probably still contain pollutants. That's because the distribution pipes themselves introduce pollutants into the water. For example, a large percentage of water in the United States is distributed via asbestos-cement pipes. Microscopic asbestos fibers, which are known carcinogens, enter the treated water from these pipes.

Another type of commonly used distribution pipe is polybutylene, or PB pipe. (PB plastic pipe is easily identified by its distinctive gray color.) PB pipe is porous to toxic solvents such as gasoline and paint thinner, and to many common pesticides and herbicides. Since PB pipe is often used to connect residences to water mains, chemicals used outdoors for gardening and maintenance can pass through the walls of the pipe and enter the water supply coming into a house.

Lead, cadmium, and other toxic metals leach out of valves and pipe couplings between water treatment plants and faucets in the home. So even if your water utility carefully monitors its water and is in compliance with government standards, there is a strong probability that the water coming out of your faucet will have picked up additional pollutants along the way.

● What's Likely to Be in Water in the Wild

Crystal clear, rushing streams; plunging waterfalls; bubbling brooks—all convey an image of freshness and purity. But when you drink from them, you are not the only drinker. Insects drink, too, and some of them lay their millions of eggs in the water. And animals drink, too, and often leave behind their wastes. One swallow of that fresh-looking (and fresh-tasting) water contains literally billions of microorganisms, some harmful and some harmless to humans. Even cold, swiftly flowing water contains bacteria, viruses, and, quite possibly, microscopic parasites and worms. The chance of getting sick, sometimes seriously sick, just isn't worth it unless it's an emergency and you have no other source of liquid.

There are several compact water filters designed for use in the wild, and I strongly recommend that you use one if you don't carry your own water with you. (These products are described in chapter 12.) Alternatively, boil the water for five minutes before you drink it.

● What's Likely to Be in the Water of Your Region

It's difficult to generalize about water quality in geographical regions because water sources vary from community to community, and often a given community will have more than one water source. Sometimes a water source is far distant from the community it serves. In the case of San Francisco, its primary water source is in

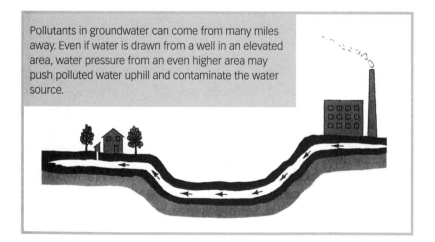

Pollutants in groundwater can come from many miles away. Even if water is drawn from a well in an elevated area, water pressure from an even higher area may push polluted water uphill and contaminate the water source.

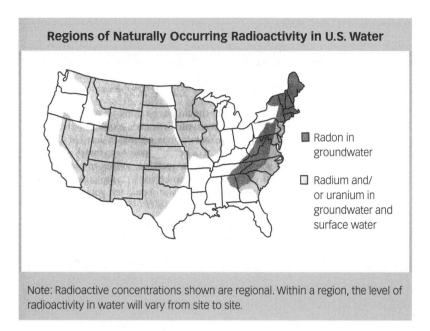

Regions of Naturally Occurring Radioactivity in U.S. Water

■ Radon in groundwater

☐ Radium and/or uranium in groundwater and surface water

Note: Radioactive concentrations shown are regional. Within a region, the level of radioactivity in water will vary from site to site.

the Sierra Nevada mountains near Yosemite National Park, some 150 miles away.

Groundwater sometimes moves great distances through rock strata, so pollutants from a source many miles away can end up in your water. Even if there are no apparent sources of pollution near the location of your water supply, that is not a guarantee that your water is free of pollutants.

With private water systems, water quality varies greatly from house to house within an area. For example, in locations where the EPA has tested wells for the presence of radon, it has been found that one house can have extremely high levels while the adjacent house has none. So if you hear that water in the middle of Wyoming, for example, is very pure, don't assume that your water is pure just because you live in the middle of Wyoming.

The light gray areas in the map of naturally occurring radioactivity in the continental United States shows where there are relatively high levels of radium and/or uranium in groundwater. If you live within one of these areas, it doesn't necessarily mean your tap water has dangerous levels of radioactivity, but you should be aware of the possibility. Radon mapping of the United States is less complete, but in general the ground formations in the dark gray area extending from Alabama up to Maine are known to have high concentrations of radon. If you live within this area, be aware of the possibility that high concentrations of radon may exist in your water or in the air inside your house.

◆ Pollutants from within Your Home

Even if water could be delivered to your home completely free of all pollutants, it probably would not be pure when it came out of your faucet. This is because household pipes and plumbing fixtures almost always add some pollutants to the water. The most notorious household source of water pollution is your faucets. Older faucets especially, are often made from a brass alloy that contains some lead. When water remains in contact with the inner walls of a faucet for several hours—overnight, for example—some lead leaches into the water from the metal. If the faucet isn't first used for wash water the next morning, whoever drinks the first glass of water may get a dose of toxic lead.

Residential water heaters are made from several metals and various compounds for seals, gaskets, and so on. Leaching of chemicals from these materials is greatly increased by high temperatures. So even though it is quicker and more efficient to use hot water for cooking, your cold water will have less impurities and should be used if possible.

◆ Indirect Health Risks

If your water comes from an underground source and the water isn't aerated, whether the source is your own private system or a water utility company, it's important to find out if there's radon in the water. If there is, that radon will be a health threat (from inhalation) wherever water is used in your house. If your water is from a utility, their test report will show whether or not radon has been detected.

If your tap water comes from a surface source or from an underground source that's aerated during treatment, you don't need to test your water for radon. If your water is from a utility, they will be able to give you this information. Note, however, that the primary danger from radon is from inhalation. The health risk from naturally occurring, underground radon gas seeping into the air in your house is much greater than the risk from radon in the water. For further information on this, see appendix A.

Most water sources, but especially groundwater sources, contain some amount of volatile organic chemicals known as VOCs. Some VOCs in drinking water are toxic if ingested but pose another threat, too. VOCs present in water evaporate readily when exposed to air, especially when the water is hot, as in showers or baths or when cooking. Once these chemicals have evaporated, inhaling them or absorbing them through the skin also may pose a health threat. If your water is from a city or small town water utility and their reports indicate that no VOCs have been detected, you don't have to test your own tap water for them. If VOCs have been detected, especially if any are above the established limit, then you should test your own tap water for them. They may have dissipated between the water treatment plant and your faucet, but they may still be in your tap water.

Testing Your Water

Is it possible to know for sure what's in your water without having it tested? No. Is it possible to find out everything that's in your water if you do have it tested? Again, no. What testing *can* do for you is tell you whether any of the major known pollutants are in your water. There are simply too many possible pollutants to test for all of them. For example, viruses are believed responsible for many waterborne diseases. But there are hundreds of types of viruses, and testing methods don't exist for even half of them. Nevertheless, you can get a lot of useful information by having your water tested.

◆ Bogus versus Genuine Water Tests

Some of the water tests performed in the United States are deliberately intended to mislead you. That's a strong statement, but here's why it's true: A large number of water tests are not done by test laboratories but by water conditioner dealers. Water conditioner dealers specialize in testing for and removing *nuisance pollutants* in water. Nuisance pollutants are those substances that are essentially harmless to health but which cause water to have a bad taste, smell, or appearance.

Very few water equipment dealers have the sophisticated equipment and expertise required to detect a variety of harmful pollutants. What they do have are test kits designed to produce dramatic visual effects. The most common of these is the mineral precipitation test. All tap water contains minerals in solution—that is, minerals that are dissolved and thus they can't be seen. When something causes them to come out of solution (in water chemistry terms, to precipitate), they become visible particles, usually white- or brown-colored.

What the water conditioner salesperson does is add a chemical to your tap water that causes the minerals to come out of solution, and presto!—there's a heap of strange-looking stuff in the bottom of the glass. This is supposed to demonstrate that your tap water is full of harmful substances and that you need an expensive water purifier or water softener to get rid of them. But the substances are really harmless and may even be beneficial minerals, such as calcium and magnesium. Everyone in the industry knows about these "tests" and many legitimate dealers condemn them; but they're still being used to deceive customers.

In fairness, the majority of water equipment dealers do offer legitimate tests for nuisance pollutants and some potentially harmful minerals, but they simply don't have the capacity to test for the majority of harmful pollutants. That's what water testing laboratories are for.

◆ What You Need to Know about
Water Testing Laboratories

If you decide to have your water tested, be careful in choosing a test lab. It's like choosing a doctor, plumber, or auto mechanic—some are better than others. Before you order tests from any lab, contact your state certification officer to see if it is state-certified for the tests that you need. (See Water Testing Laboratories, in appendix C for an EPA website with links to information on state certification offices.)

An important thing to understand here is that lab test results are limited by the smallest amount of pollutant that a lab can detect. So whenever amounts of pollutants are below those thresholds, it doesn't mean that those pollutants aren't in your water; it means that the lab didn't detect them at the sensitivity level of their equipment. The clue to this is to watch for the "less than" sign (<). This sign indicates that the number given is the lab's lowest detection limit.

When you receive test results from a water testing lab, or from your water utility, you will see results for a given pollutant expressed using one of four measures:

- mg/l means milligrams per liter of water
- ppm means parts per million
- mg/l means micrograms per liter of water
- ppb means parts per billion

Milligrams per liter and parts per million are almost identical and are often used interchangeably. Similarly, micrograms per liter and parts per billion are almost identical and are used interchangeably.

Maximum amounts of some pollutants allowed in water by government standards may be higher than the level that would ensure the water is safe. One reason is that many water utilities would be forced to take expensive measures to remove pollutants down to a lower, more stringent limit. And some water utilities, especially the smaller ones, simply couldn't afford to comply.

When water-testing laboratories give you a report on your water, they'll point out any pollutants in your water that exceed government standards. That's certainly helpful information, but keep in mind that if pollutants are present, they may be harmful even at levels below the allowable limits. Nevertheless, testing can be very valuable and it is necessary in some cases.

Testing your water can cost you a lot of money if you don't go about it correctly. The majority of water testing laboratories are not efficiently set up for the general public. They do most of their work for cities, counties, water utilities, and commercial customers. When you have your water tested by one of these labs, it's like buying a pair of handmade shoes—good quality but extravagantly expensive.

If you wanted to test for most of the possible pollutants in your water, the cost from a conventional lab would run in the thousands of dollars. Additionally, water quality can change with the seasons, due to rainfall, because of agricultural runoff, and so on, so it might be necessary to retest your water. All told, getting a good picture of what is consistently in your water from a conventional laboratory would probably cost you $10,000 to $20,000!

Fortunately, there are alternatives. In recent years, to meet consumers' needs for a broad range of water tests, several labs have developed sophisticated automated testing procedures. They offer comprehensive tests for relatively low prices. When you order a test from them, you receive a special mailing package. You place your water sample in the mailer, send it to the lab, and receive your report in a couple of weeks. With automated labs such as these, you'll pay about $90 for a fairly comprehensive test series, and up to about $300 for a complete battery of tests. (See Water Testing Laboratories in appendix C for a recommended comprehensive test.)

◆ Tests You Can Do Yourself

Self-testing kits are available at a relatively low cost. Though they're more limited than laboratory tests, they can be useful in certain situations. If you have evidence that a particular pollutant is or has been in your tap water, self-tests may allow you to determine if it continues to be there. You may want to test for that pollutant several times over the course of a year to allow for seasonal fluctuations. If the pollutant in question is bacteria or a common toxic metal, you can buy inexpensive and easy-to-use test kits. However, you cannot self-test for viruses or parasites, organic chemicals, rare metals, or radioactive substances. (See Water Testing Laboratories in appendix C for links to companies that sell self-test kits.)

So . . . Should You Test Your Tap Water? A Summary

This chapter has presented a lot of technical information, and it may be hard to assimilate all of it. So I'll summarize when testing is appropriate and what kinds of tests to choose.

◆ When Your Water Is from a Utility

- Get your utility's test report and see which, if any, pollutants are above government limits. If none are above the limits, you probably don't need to have further tests done. If the utility's test report does show any pollutants above government limits, make a note of them for future reference.
- If the utility's report shows the presence of *any amounts of* VOCs, have your tap water tested for their presence.
- Whether or not the utility's report shows the presence of radon, check with your county health department to see if high concentrations of radon have been detected in your area. If the answer is yes, have your house tested for *airborne* radon (and read appendix A).
- Remember that there will be pollutants in your tap water that don't appear on the utility's test report, either because the test doesn't cover them or because they're present at levels too low to be detected. You'll still need an alternative to drinking tap water.

● When Your Water Is from a Private Well or Spring

- Get a test for bacteria at least once each year, preferably during the rainy season.
- Get a comprehensive test for pollutants at least once every other year. If any pollutants are shown to be above government limits, make a note of which ones.
- If the comprehensive test shows VOCs to be in excess of government limits, consider using a whole-house water purification system.
- Check with your county health department to see if high concentrations of radon have been detected in your area. If the answer is yes, have your house tested for *airborne* radon (and read appendix A).
- Remember that there may be pollutants in your tap water that don't appear on the comprehensive test report, either because the test doesn't cover them or because they're present at levels too low to be detected. You'll still need an alternative to drinking tap water.

● Tap Water Profile

Keep all the information you gather to make a profile of your tap water. If you decide to use a water purifier now or in the future, the profile will help you decide which kind to get.

Tap Water Profile		
	Type of Pollutant	**Name of Pollutant(s)**
Pathogens	Bacteria	
	Viruses	
	Cysts and other parasites	
Minerals	Toxic metals	
	Nitrates and other nonmetals	
	Asbestos fibers	
Organics	Volatiles (VOCs)	
	Pesticides, PCBs, THMs, herbicides, and other nonvolatiles	
Radioactives	Radon	
	Uranium and radium, dissolved	
	Uranium and radium, particles	
Additives	Chlorine	
	Fluoride	
	Flocculants, alkalizers, and other water treatment chemicals	
	Organic additives	
Tastes and Smells	Hydrogen sulfide and other volatiles	
	Dissolved minerals	
	Mineral and organic particles	

CHAPTER 4

The Best Drinking Water for Good Health

Chapters 2 and 3 described the different types of toxic pollutants that can contaminate your water—pollutants that are health hazards. Given all this information, it might seem reasonable to conclude that the purest water—water that is 100 percent H_2O—is the best drinking water. But not all the substances in water are necessarily harmful to your health. In fact, some substances in water are beneficial.

Distilled Water

There has been much debate over the years about distilled water. Advocates claim it's a cleansing agent that will purify your body and leave you in glowing health, while detractors sound the alarm that it will rob your body of minerals and leave you a quaking skeleton. So what is distilled water, anyway, and how does it affect you?

Distilled water is water that has had essentially all of the dissolved substances within it removed by evaporating it and condensing it back to liquid form. Since about 99 percent of all dissolved matter in water is minerals, distilled water is essentially water that has had all of its minerals (as well as pollutants) removed. The information presented here on distilled water also applies to water that has been demineralized by other methods and is known as *purified, demineralized*, or *deionized* water. I use the terms *distilled water* and *demineralized water* interchangeably.

The arguments in favor of drinking distilled water claim that the minerals present in water clog the body and impede bodily functions. This theory is partly based on the notion that since calcium (the primary mineral in water) helps to form bones and to harden things in general, it must harden other things in your body as well. Literally dozens of books written by well-meaning doctors, nutritionists, and practitioners of holistic medicine claim that essentially all disease is caused by minerals in water and that by drinking distilled water you will be reinvigorated, your arthritis or hardened arteries will disappear, and so on. These well-meaning authors apparently know very little about water chemistry, and while they present anecdotal "evidence" to support their theories, there have been no valid scientific studies to verify them.

On the other hand, one of the arguments against drinking distilled water is that you lose a primary source of necessary minerals in your diet and, further, that because the water has lost its own minerals, it attracts and grabs minerals within your body, causing a mineral deficit. As with the previous argument, there is no evidence to support this claim.

In fact, all naturally occurring water, even rainfall, contains minerals. Distilled water, having lost its minerals, is in an unnatural state. In chemistry terms, distilled water is aggressive. That means that it tries to regain minerals to bring it back to the water's natural state.

When you drink distilled water, it undoubtedly attracts some minerals from whatever surfaces it contacts, meaning your mouth, throat, and stomach. But the amount of minerals it needs is far less than the amount naturally present in these surfaces. By the time it reaches your stomach, it has regained enough mineral content that it's no longer aggressive.

Here are the facts: We get most of the minerals we need from fruits and vegetables, whose minerals are in a form that are more readily digested than waterborne minerals. You're unlikely to experience mineral deficiencies if you eat a normal diet including a variety of foods. At the same time, studies have shown that highly mineralized water can have beneficial effects on health. I'll cover this point later in the chapter.

Some people don't like the taste of distilled water, which is sometimes described as tasting flat. Water bottling companies know that demineralized water generally doesn't taste as good, and that's why they sometimes add minerals to their bottled water after it's been demineralized during the purification process (see chapter 6 for more information on this). Such "remineralized" waters are usually labeled "drinking water."

As I mentioned, distilled water is aggressive, and while it won't hurt you to contribute some minerals from your own body to neutralize it, storing distilled water is another matter. Remember that water in its natural state always contains minerals. Since distilled water is in an unnatural state, lacking its usual complement of minerals, it tries to correct this by combining with any other substances it can. When it's stored, the only substances available for it to combine with are the things the water contacts—containers and if the containers are open, air. But since containers are usually closed and air is relatively free of harmful pollutants, containers are really the only potential problem.

No container is completely inert (chemically inactive). Because of its aggressiveness, distilled water will leach out some of whatever the container is made of. Glass is the most inert material for containers, and therefore the safest. However, distilled water is routinely sold in plastic bottles even though there's a tendency for the plastic to enter the water over a period of time. Although results are inconclusive, some studies suggest that trace amounts of the plastic from bottles may be harmful in the long term. (Chapter 6 covers the relative safety of different kinds of water containers.)

I recommend that distilled water, whether purchased in bulk or produced at home by your own distiller, be stored in glass bottles as soon as possible. And if you drink distilled water regularly, make sure that your diet contains an assortment of mineral-rich fruits and vegetables. This is good advice, anyway, regardless of what type of water you drink.

Hard versus Soft Water

Hard water contains large amounts of dissolved calcium and, to a lesser extent, magnesium. *Soft water* contains relatively small amounts of calcium and magnesium. Water can be naturally hard or naturally soft. The disadvantage of hard water is that more soap or detergent is needed to get clothing, dishes, and other items clean. Also, soft water makes skin feel smoother and hair feel softer. Soft water has another advantage: It causes less scale, the hard, whitish stuff that forms on the insides of pipes and tends to clog up the innards of water appliances. A huge industry exists in the United States whose sole purpose is to make hard water soft. There are hundreds of water softener manufacturers and tens of thousands of water softener dealers.

Although soft water does occur naturally, most of the soft water consumed in this country is created by water softeners. These simple gadgets trap calcium and magnesium in the water and replace it with salt (sodium chloride). So when you drink softened water, you're simply drinking water with most of its calcium and magnesium removed and with salt added.

The higher the levels of calcium and magnesium in tap water, the more salt is exchanged in a water softener. If your incoming tap water has a low level of minerals, your softened water will be low in salt. If your tap water has a high level of minerals, your softened water will have a high level of salt.

There has been much publicity over the years about the negative health effects of drinking softened water. This springs from some early research that showed a correlation between cardiovascular disease and high-salt diets. But more-recent research has shown that the amount of salt consumed by drinking softened water

Calcium and magnesium in water create hardness.

A water softener exchanges salt for calcium and magnesium.

The more calcium and magnesium in the incoming water, the more salt there will be in the softened water.

The softened water is low in calcium and magnesium and higher in salt.

is insignificant when compared to overall daily salt intake. Only people who are on a severely salt-restricted diet face a health risk from the added salt in softened water.

In recent years, a substitute for salt (sodium chloride) has become popular in water softeners. This substitute is potassium (potassium chloride). When potassium is used, the calcium and magnesium in hard water are exchanged for potassium, so no sodium is added to the water. Potassium, in general, is a vital, health-promoting nutrient, but some health conditions do require that people follow potassium-restricted diets.

The Benefits of Minerals in Water

Although we get the majority of the minerals in our diet from fruits and vegetables, minerals in water do play a role in the maintenance of good health. Over the past thirty-five years, research has continued to amass in support of the health benefits of minerals in water. Studies of populations in areas with naturally occurring hard water (high mineral content) and soft water (low mineral content) have found fewer occurrences of cardiovascular diseases, cancer, diabetes, respiratory diseases, and other health problems that occur in areas with hard water.

Almost all of the minerals in water are dissolved. That is, they have liquefied and merged with the water. The standard measure of dissolved minerals in water is called TDS, which stands for *total dissolved solids*. Studies show that disease is statistically less prevalent in areas that have moderate to high TDS levels. However, high TDS levels don't seem to provide any advantage over moderate levels; areas with high and moderate TDS levels show the same health improvements when compared to areas with low TDS.

What this suggests is that once an adequate minimum intake of beneficial minerals from water is achieved, ingesting higher levels doesn't confer additional benefits.

And whether you drink water with lots of minerals or distilled water, remember that most of your minerals come from fruits and vegetables, so the amount of these in your diet is more important than the level of minerals in the water you drink.

In the process of removing harmful pollutants from water, beneficial minerals are sometimes removed as well. In general, when pollutants are present, it's more important to remove them than it is to save minerals.

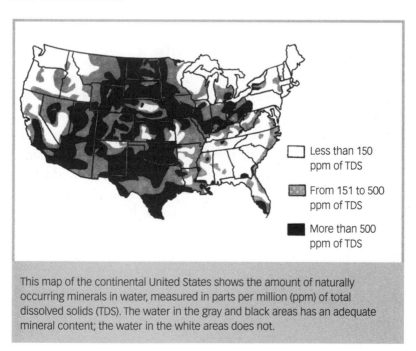

Less than 150 ppm of TDS

From 151 to 500 ppm of TDS

More than 500 ppm of TDS

This map of the continental United States shows the amount of naturally occurring minerals in water, measured in parts per million (ppm) of total dissolved solids (TDS). The water in the gray and black areas has an adequate mineral content; the water in the white areas does not.

Water from Natural Sources

Dating back to the first miracle cures at Lourdes, France, in 1858, spas, springs and deep aquifers have enjoyed the reputation of producing superior (health-promoting and in some cases disease-curing) drinking waters. Unlike spas (natural mineral springs) in the United States, those in Europe are closely regulated and any potential sources of pollution are prevented by law from operating near them. In spite of this, there have been several instances of toxic pollutants discovered in spot checks of European spa water. The likely source of the pollution is often unknown, but it's thought that the chemicals used to sterilize bottling machinery may be the cause. The discovery of pollutants in spa water has occurred enough times to suggest that these are not isolated incidents but rather a recurring problem. I can't vouch one way or another for the healing powers of natural-source waters, but pollutants have also been found in some natural-source waters in the United States, and I recommend that you take certain precautions, described in chapter 6, before drinking water from natural sources on a regular basis.

Health Claims for Exotic Waters

In recent years, several novel water treatment products have appeared on the market with claims to promote better health and to energize and purify the body by adding oxygen to water or by changing the molecular structure of water. These products are said to produce health-promoting water that is known by such names as hexagonal water, vortex water, magnetic water, ionized water, and polyatomic oxygen water. It is beyond the scope of this book to investigate and report on the efficacy of these waters, but for readers who are interested, I have included some references on this topic in appendix C.

How Much Water Should You Drink?

Many articles have been written over the years about the detrimental health effects of not drinking enough water. Other articles claim that drinking beverages other than water is not only inferior to drinking water, but that beverages other than water act as dehydrating agents and rob the body of water. Still other articles base health claims on when you should drink water (before meals, after meals, mornings, evenings, and so on) and even how you should drink water (through a straw rather than from a glass or bottle).

It's beyond the scope of this book to go into all of these topics. Suffice it to say that as of this writing I have not found any credible, convincing studies to indicate that coffee, tea, juice, milk, soda, or other liquids contribute any less to the body's water supply than drinking plain water. However, various studies do show that every aspect of bodily function can be impaired by insufficient consumption of liquids. In general, when you drink an adequate amount of liquids, your body will be better able to neutralize and eliminate harmful substances. An interesting aside is that insufficient consumption of liquids causes the body to *gain* weight. This is because fat cannot be completely metabolized, or converted to energy, without adequate water in the body.

So what is an adequate amount of liquid? Should you, according to the old maxim, drink eight glasses of water per day? Recent studies indicate that this may be too much for some individuals. The consensus is that your body will tell you when it needs liquid. When you are thirsty, drink; when you are not, don't.

The Message from Water Crystals

Dr. Masaru Emoto, a Japanese water scientist, has used high-speed photography to capture images of the formation of water crystals as water begins to freeze. His initial discovery was that the crystals of pure, unpolluted water formed intricate, beautiful, symmetrical patterns, while the crystals of polluted water or tap water were incomplete and distorted. But then came the real shockers: Various *written* notes which had been left in close proximity to the water being photographed, appeared to have had an effect on the formation of crystals. Repeated experiments indicated that when the water being photographed was in proximity to positive words such as *love* and *gratitude*, they produced beautiful crystals, even if the water was polluted. In contrast, negative words such as *anger, irritation,* and *frustration* produced poor-quality crystals, even with pure water. Emoto's next step was to experiment with thought processes, and indeed, the experimenters' positive thoughts produced attractive, well-formed crystals, while negative thoughts had the opposite effect.

Dr. Emoto believes that water is a universal carrier and reservoir of the thought processes of all human beings in the world. He's published several books on his work, including *Messages from Water* and *The Hidden Messages in Water,* both of which have become best sellers. Whether you believe this field of inquiry has any validity or you dismiss it out of hand, it seems reasonable to suggest that whatever kind of water you drink, think good thoughts when you drink it—it can't hurt. (See Unproven or Untested Water Concepts in appendix C for a link to a critical view of this subject.)

Water for Good Health—A Summary

- It's more important to remove harmful pollutants from water than to save beneficial minerals that may be in it.
- If you drink distilled water, be sure to include a wide variety of fruits and vegetables in your diet.
- Don't store distilled water in plastic bottles; use glass.
- All other things being equal, hard water is healthier than soft water.
- And again, an important reminder: Don't drink chlorinated water supplied by a water utility on a regular basis.

CHAPTER 5

Four Simple Ways to Improve Tap Water without Special Equipment

If you have to drink tap water (and there are times when we all have to), here are a few tips on how to reduce some of the pollutants that may be in it.

♦ The Twenty-Second Wait

When water sits motionless in pipes overnight, there's much more time for toxic chemicals to leach into the water from pipes and plumbing fixtures. The most notorious culprits are older faucets, many of which contain lead. Water that's been in a faucet overnight or longer will be the most heavily polluted, so when you draw water from a tap that has not been used for a while, allow the water to flow for about twenty seconds before using it for drinking or cooking.

Even when the water has not been standing for a long time, the change in water pressure when you turn on a faucet suddenly can cause pollutants to break loose from pipes and fixtures. So when you're drawing tap water for drinking or cooking, try to remember to turn the water on gradually, then wait about ten seconds before filling your container.

● Open Air

When water from a public water supply comes out of the tap, it still has residual chlorine in it from the disinfection process. Chlorine in water is volatile; that is, it evaporates easily. Just allowing your tap water to stand in an open container with a wide mouth, such as a cooking pot, for a few hours will allow most of the residual chlorine to evaporate.

● Boiling

Boiling water for ten minutes is an excellent way to disinfect it, and boiling also removes any remaining chlorine or other volatile pollutants that may be in the water. There has been some criticism of boiling tap water because, it's claimed, nonvolatile pollutants become more concentrated as the level of water decreases during boiling. This is not a valid argument. The amount of water evaporated during ten minutes of boiling is quite small, certainly not enough to significantly concentrate any remaining pollutants.

If you're going to boil your drinking water, use a stainless steel, glass, or porcelain pot. These are the safest utensils. For many years there has been suspicion of a link between cooking in aluminum pots and pans and Alzheimer's Disease. However, recent studies indicate that while there is a link between *naturally occurring* aluminum in water and Alzheimer's, the aluminum molecules released from cooking utensils are of a type that do not affect the body.

● Stirring

Stirring tap water in an electric blender at low speed (to avoid splashing) for ten minutes will cause volatile contaminants to evaporate, as in boiling. If you use this method, be sure to do it with the blender cover off to allow for evaporation.

Keep in mind that although the above methods will improve tap water, they can't remove all of the potential pollutants. If you'll be storing the water after using any of these methods, refrigerate it to minimize possible growth of microorganisms. Tap water treated by these methods should only be used in place of adequately treated water for short periods of time.

CHAPTER 6

Bottled Water, Vended Water, Bulk Water Stores

Bottled Water

As for other aspects of water, the arguments for and against bottled water can be very heated. Opponents claim that the source for most bottled water is the same as your tap water (if your tap water is from a water utility), but they fail to understand that bottled water undergoes additional treatment after processing by the water utility company. On the other hand, there can be wide variations in the quality of bottled water from one type or brand to another.

A good place to begin sorting out the pros and cons is with a description of how bottled water is regulated. In theory, bottled water is regulated by the Food and Drug Administration (FDA) and by each state. In FDA terms, all bottled water is legally defined as a beverage. While water utilities delivering your tap water are regulated by the Environmental Protection Agency (EPA), all beverages are regulated by the FDA. The FDA regulations for pollutants in bottled water generally follow those established by the EPA but in addition require all bottled water companies to adhere to stringent sanitary practices in the bottling process. Although, in theory, regulation of bottled water by the FDA is similar to regulation of tap water by the EPA, in practice the FDA has consistently dragged its feet on establishing standards for bottled water—standards that it is legally required to establish. And the FDA's enforcement of those standards it has established are weak at best.

Among the many weaknesses of FDA regulation is the fact that the FDA exempts intrastate bottled water (bottled water not sold outside the state where it originates) from regulation, even though 60 to 70 percent of all bottled water sold in the United States is intrastate. Seltzer, other carbonated waters, and flavored waters are also exempt. The FDA requires testing of bottled water less frequently than is required by the EPA for tap water, and if a particular bottled water exceeds the limits for any pollutant, that water can still be legally sold as long as it includes on its label a statement such as "contains excessive chemical substances."

For these and several other reasons, FDA regulation of bottled water is inadequate.

The second level of bottled water regulation is by each state. The main responsibilities of the states are to inspect bottling facilities, monitor water quality, and make sure that both FDA and state regulations are adhered to. But states have no legal obligation to comply with FDA requirements, and they are free to establish (or not to establish) their own bottled water monitoring systems.

Some states have developed relatively strong programs for bottled water. Among these are California, Florida, Louisiana, Maine, Maryland, Massachusetts, Montana, Nevada, New Jersey, Texas, Vermont, Washington, West Virginia, and Wisconsin. Some of the weakest or nonexistent bottled water programs are in Arizona, Delaware, Illinois, Indiana, Missouri, North Dakota, Utah, and Virginia. In general, most state programs have no requirement for informing the public when pollutants in bottled water exceed FDA or state limits.

The third level of regulation of bottled water is by the International Bottled Water Association (IBWA), a private trade group that certifies that the bottled water of all its member companies meets considerably stricter standards for purity than the FDA and state requirements. This is a good program that is superior to either FDA or state requirements. The actual certifications for the IBWA are performed by NSF International, a highly respected and accredited testing and monitoring service. The drawback is that around 15 percent of water bottlers are not members of the IBWA and do not participate in this program. (See Bottled Water Information in appendix C for a link to a list of member companies.)

There are six basic types of bottled water: purified water, drinking water, fluoridated water, mineral water, natural-source water, and specialty water. The source water for the first three types is municipal water, the same water that comes out of your tap if you get your water from a water utility.

But these three kinds of bottled water then undergo another treatment process that further purifies the water. The following illustration shows how typical bottling plants treat water after it has already been treated by the water utility.

As shown, a bottling company will typically filter the water to remove any remaining dirt or particulate matter; demineralize it to remove any remaining toxic minerals and chemicals; and aerate it to remove any odors by exposing it to air. After having its mineral content largely removed, water tastes flat, so for drinking and fluori-

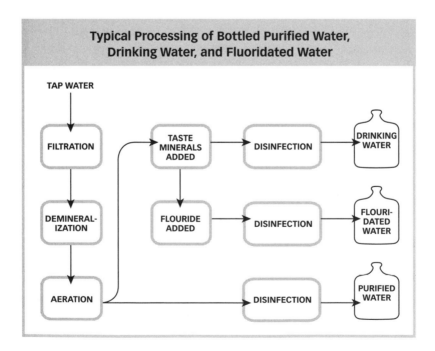

Typical Processing of Bottled Purified Water, Drinking Water, and Fluoridated Water

TAP WATER

FILTRATION → TASTE MINERALS ADDED → DISINFECTION → DRINKING WATER

DEMINERAL-IZATION → FLOURIDE ADDED → DISINFECTION → FLOURI-DATED WATER

AERATION → DISINFECTION → PURIFIED WATER

dated water some safe minerals are added to improve taste. Fluoride is only added to fluoridated water. As the final step before bottling, all three types are disinfected by ozone or ultraviolet light.

◆ Purified Water

Purified water may also be labeled "distilled water," "water for distillation uses," "deionized water," or "demineralized water." Each of these labels indicates that, in addition to having gone through other cleansing processes, the water has had essentially all of its minerals removed and none added back in. So, in the sense of there being nothing in it but H_2O, this is the purest water. However, more relevant than how pure it is, is whether it's good or bad for you.

As explained in chapter 4, water without minerals will neither harm you (by supposedly draining your body of necessary minerals) or help you (by supposedly cleaning out your body). However, there is a problem specific to *bottled* purified water. Purified water has had essentially all of its nonwater constituents (which are mostly minerals) removed. This is not a natural state. In its natural state, water contains many different kinds of minerals. Once purified, water will try to return to its natural state by reacting chemically with any material it contacts. Because of this tendency, purified water is known as *aggressive* water.

When purified water is placed in soft plastic bottles, it tends to leach out any chemicals in the plastic that are loosely bonded. These chemicals then enter the water in miniscule amounts. The longer purified water remains in a plastic bottle and the higher its temperature, the greater the chance of chemicals entering the water. While there is no strong evidence to prove that this is harmful, the long-term effect of trace amounts of these chemicals is unknown. Some studies have shown a link between chemicals leached from plastic water bottles and disorders of the human immune system, but this connection is far from conclusive.

In chapter 2, I described the concept of overall health risk and recommended that you try to reduce your overall toxic load as much as is practical. For this reason alone, it seems reasonable to avoid another potential source of pollution by avoiding contamination from plastic containers. Note, however, that this applies only to purified bottled water. To repeat, this is bottled water with any of the following labels: "purified water," "distilled water," "water for distillation uses," "deionized water," or "demineralized water."

Recommendation: Don't buy purified water in soft plastic bottles.

● Drinking Water

Bottled water labeled "drinking water" is either water that has been partially purified but retains most of its original minerals or water that has been thoroughly purified (completely demineralized) and then has had minerals added to improve the taste. Since replacing minerals eliminates most of the aggressiveness of purified water, there is less likelihood of chemicals leaching from soft plastic containers into this type of water.

● Fluoridated Water

The subject of fluoridation in general is covered in chapter 2. In terms of bottled water, fluoridated water is identical to drinking water except that fluoride has been added.

● Mineral Water

Mineral water can be derived from any source, municipal tap water or water from a natural source. It can be labeled "mineral water" as long as it has a certain level of dissolved minerals in it. The FDA's legal minimum is 250 parts per million (ppm) of total dissolved solids (TDS). This minimum may be naturally occurring, or the bottler can add minerals to tap water. Because of the uncertainty of the source of this type of bottled water, I recommend that it not be used; note the following information on natural-source water.

● Natural-Source Water

Natural-source bottled water includes all those water products whose source is claimed to be a naturally occurring spring or aquifer (underground reservoir). Most natural-source water is healthy and safe, and it generally contains high levels of beneficial minerals. But as mentioned in chapter 4, the problem with these products is that there's no guarantee they're healthy or safe: Water from some natural sources has been found to have abnormally high levels of naturally occurring radioactivity. Some spring waters have been shown to contain high amounts of toxic heavy metals, and others, even "prestige" brands of bottled water, have been shown to be contaminated with toxic organic chemicals. On the other hand, most bottlers of natural-source water do regularly test their water for pollutants, and this type of bottled water is regulated by the FDA, individual states, and, for member companies, by the IBWA.

◆ Specialty Water

Specialty water includes products with added flavors and/or carbonation, such as seltzer or lemon-flavored sparkling water. Some of these products use water from naturally occurring springs and some use tap water. Usually sold in small containers, these specialty waters are legally categorized as beverages—that is, it is assumed that, like beer or juice, they will be consumed only occasionally and not as a daily staple. Specialty bottled waters are exempt from regulation by the FDA and are probably no more or no less safe than other beverages—which isn't saying a lot.

◆ Recommendations for Bottled Water

- For all of the reasons detailed above, of the six basic kinds of bottled water (drinking water, purified water, fluoridated water, mineral water, natural-source water, and specialty water), drinking water and purified water are the best overall choices, with the condition that purified water is not stored in plastic bottles.
- Choose a well-known, major brand of water; the large companies generally employ the most effective purification processes and have the best quality control. Try to choose a brand produced by a company that's a member

of the IBWA, and thus subject to more stringent testing.
- Buy bottled water from a store that sells a lot of water, and thus has the water on its shelves for the shortest time.

Store bottled water in a cool place. Never allow stored water to be in direct sunlight. Sunlight increases some microbial growth and allows more chemicals from the container to enter the water.

Don't drink directly from a water bottle unless you're going to finish the bottle within a day or two; you're transferring millions of microorganisms from your mouth to the water each time you do this.

Vending Machine Water

If you can put up with the inconvenience of lugging your bottles to and from a machine, you should seriously consider vending machine water as an alternative to commercially bottled water. When a vending machine is properly maintained, the water it produces is as good as high-quality bottled water.

Vending machine water always uses municipal tap water as its source. And while municipal tap water is not of high quality, it is usually quite consistent. Because of this, and because the purification processes in the machines are proven and effective, the quality of water they produce is predictable. In fact, the biggest problem with using vending machine water is usually not the machine itself but the cleanliness of the containers you use to collect the water.

Each state sets up its own standards for water vending machines. They regulate how the machines must be constructed and what purification methods are used and also set minimum standards for the quality of water produced. County agencies are responsible for enforcing the state standards. Sometimes health inspectors wear two hats—those of both county and state inspectors at the same time.

If you're going to use water from vending machines, you need to be willing to do a little homework. The first thing you should look for when you consider using a water vending machine is a seal or label of certification by your state or county. This means, at least in theory, that a particular machine is regularly maintained and inspected. When you find this label on the machine, call your local health department. Make sure you have the location of the machine written down or in mind. Find out which county agency is responsible for inspecting it and how the agency determines that the machine is being properly serviced. Keep at it until you find out for sure that the machine is being properly maintained. If you run into a dead end with the county, or if you can't find a label of certification on the machine, look for the manufacturer's label, which is legally required to be on it. Take down their telephone number and call them. Tell them the location of the machine you want to use and ask them how they service it. If after all your efforts you still can't verify that the machine is properly maintained, don't use it.

Typical vending machines have several stages of purification, as shown in the illustration on the following page.

In the first stage, a sediment filter traps any dirt or other particles that may be in the water. The second stage of purification is reverse osmosis. This process employs a thin synthetic membrane with ultratiny openings the size of small molecules; these trap chemical pollutants but allow water to pass through. Most pollutants and most, but not all, of the harmless minerals are removed by this process.

Many vending machines allow you to choose between drinking water and purified water (the latter may also be labeled "distilled," "deionized," or "demineralized" water). The only difference between the two is that purified water goes through the demineralization stage and drinking water doesn't. When purified water is selected, the demineralization process removes all remaining minerals from the water. This makes the water suitable for use in steam irons, batteries, and other applications where demineralized water is required.

Typical vending machines have separate storage tanks for drinking water and purified water. When money is deposited in the machine and the type of water is selected, the water in one or the other of these storage tanks goes through a carbon filter that removes any remaining traces of chemical pollutants and gives the water a fresh,

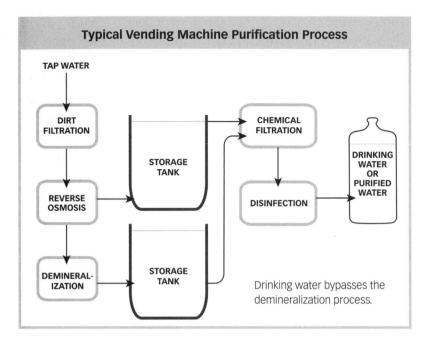

Typical Vending Machine Purification Process

TAP WATER

DIRT FILTRATION

REVERSE OSMOSIS

DEMINERAL-IZATION

STORAGE TANK

STORAGE TANK

CHEMICAL FILTRATION

DISINFECTION

DRINKING WATER OR PURIFIED WATER

Drinking water bypasses the demineralization process.

clean taste. The final stage is disinfection, which kills any microorganisms that may be in the water by means of ultraviolet light.

It's critical that the containers you use be free of pollutants. Be sure your containers have not been previously used to store any toxic or unknown substances. Wash your containers thoroughly with detergent and rinse them several times before you use them. When they aren't in use, turn them upside down after washing to let all the excess water drip out, and let all the water inside dry thoroughly before you cap the containers.

After you fill your bottles with water, cap them and store them away from sunlight and heat. The cooler the storage place, the longer your water will stay fresh. If possible, don't store water for longer than a month. When water is stored for longer periods, it starts to pick up tastes and smells and will eventually grow bacteria and other microorganisms unless it and the containers have been sterilized.

The materials your containers are made of will also affect the quality of your water. All containers, even those made of glass, leach tiny amounts of pollutants into water. In general, glass is best. Next best are containers of stainless steel, which, although they are less prone to leaching pollutants than most other metal containers, will over time leach trace amounts of toxic metals into water. Containers made of copper or aluminum can leach potentially harmful chemicals into water, so don't use them.

If you can't use glass or stainless steel and must use plastic bottles, make sure the bottles are kept in the coolest place you have available (but not subject to freezing) and away from sunlight, and keep the storage time as short as convenient.

Avoid using any plastic containers that were made for purposes other than holding food; utility containers made for nonfood use may contain vinyl compounds, which are toxic and can leach into the water. Also, don't use containers with surfaces that bacteria can grow on, such as wood or waxed paper milk cartons.

It is best to have extra containers so you can rotate them. After a bottle has been emptied of drinking water, drain out all excess water and allow the bottle to dry for a week or two, upside down and with the cap off. Drying the bottles thoroughly will kill most microorganisms that may have started to grow in them. If the water in a bottle starts to have a noticeable taste or smell, discard the water and wash the bottle thoroughly. Washing with disinfectants is also helpful, but be sure to rinse the bottle thoroughly after washing.

◆ Recommendations for Vending Machine Water

- Water vending machines generally provide high-quality drinking water if they're properly maintained. Be sure to look for the label of certification on the machine and do whatever is necessary to make sure the machine is serviced regularly.
- Read the description of the purification processes the vending machine employs. This should be stated right on the front of the machine. A few states don't require a final stage of ultraviolet disinfection. If you should find a vending machine without this final stage, it's even more important that you make sure the machine is properly serviced.

- Choose a machine in a popular location that does a lot of business. In general, the more a machine is used, the less chance there is for any standing water to grow micro-organisms. Out-of-the way machines that stand unused for several days at a time should be avoided.

Bulk Water Stores

There are two kinds of retail stores that sell bulk water: health food stores and water specialty stores. Water from health food stores is usually superior to tap water, but their purification equipment is relatively crude and may not be subject to inspection. Because it's easy to get drinking water that is of known high quality in most locations, I recommend that you do not purchase drinking water from a health food store (unless the water is from a vending machine that meets the criteria described above).

Water specialty stores, that is, stores that sell only water and water products, generally offer a very high-quality product. Because these stores specialize in water, their managers usually have a considerable knowledgeable of water purification, and the in-store equipment is quite sophisticated and is closely monitored and maintained. Like bottled water companies and water vending machines, water stores usually offer both drinking water and purified water.

Some water stores offer only purified water that has had its minerals removed by reverse osmosis or by distillation. If the water has been treated by reverse osmosis, only about 70 percent to 90 percent of its minerals have been removed. If the water has been distilled, about 97 percent to 99 percent of its minerals have been removed. As mentioned before, this is very pure water that will aggressively absorb any loosely bound materials from containers. So if you purchase purified water from bulk water stores, use glass containers.

Along with glass bottles, some water stores sell rigid, clear plastic bottles made of polycarbonate plastic; because they're more lightweight, these are generally used for delivery of water to homes and businesses. For many years they were thought to be very inert, but recent research has shown that they do leach some toxic compounds into the water over time. So even though they're lighter and less prone to breakage than glass, I recommend that you avoid using them for purified water.

All of the above information on bottled water applies to home-delivered water.

Storing Water

All water contains some live microorganisms. Typically, the microorganisms in drinking water are harmless to humans and they are antagonistic to each other—that is, they feed on each other and are self-controlling, at least in the short term. But warmth and sunlight allow microorganisms to grow faster, and in general, the longer water is allowed to stand, the higher its concentrations of microorganisms will be. For drinking water purposes, store water for as short a time as possible. However, sometimes it's necessary to store water for long periods of time. The main reason you'd want to do so is in the event of an emergency. See chapter 13 for more details on long-term storage of water, as well as information on emergency disinfection of water.

CHAPTER 7

How Water Purifiers Work

If you're considering using a home water treatment unit, this chapter will help you make an informed decision. I'll describe the basics of each type of water purifier available and discuss some of the pros and cons of each. But first, a word about "purifiers." The Environmental Protection Agency (EPA) is the federal agency responsible for setting standards for drinking water treatment products. They decided to name the units that meet their requirements for disinfecting water—that is, removing or killing pathogens—water "purifiers." So only those products that meet their disinfection standards can legally be advertised as purifiers. That leaves all the other drinking water treatment products as . . . what? "Cleaners," "improvers," or "uncontaminators"? In contrast, for purposes of simplicity, I use the term *purifier* to mean any water treatment product whose purpose is to improve water quality. (But, if you're going to buy a water treatment unit for your home, do remember that only disinfecting units can be advertised as purifiers.)

There are four basic types of water purifiers for home use:

- Filters
- Reverse osmosis units
- Distillers
- Ultraviolet units

Other methods of treating water include aeration, deionization, and ozonation. There is also a recent innovation, the air-to-water purifier, that may prove popular in the coming years.

How Filters Work

All filters use a substance that traps, absorbs, or modifies pollutants in the incoming water. The active substance within a filter is called the *medium*. There are many different kinds of filter media. Some work by mechanically trapping pollutants with an ultrafine sieve action. Others attract and capture pollutants by their electrical charge. And still others employ a process called *adsorption*, in which pollutants are trapped within the microscopic pores of the medium by chemical attraction.

All filters have one similar weakness. Have you ever noticed the slippery biofilm that develops on anything where water has been standing? Then just imagine the warm, wet inside of a filter; it's an excellent place for the growth of microorganisms. I use the word *microorganisms* here instead of harmful *pathogens* because, when tested, the biofilms within filters are made up mostly of microorganisms that cause no apparent, immediate harm. However, not much research has been done on this, so to be on the safe side and help flush out microorganisms that accumulate inside a filter, always run water through it before use. If the filter has been unused overnight, flush water through it for thirty seconds; if the filter has been unused for several days or longer, flush water through it for two to three minutes. If it's been used every few hours or more frequently, a ten-second flushing will be enough.

◆ Sediment Filters

Most sediment filters work by mechanical sieve action. They're used for removing dirt and other particles from water. If your water comes from a private well, it may contain sand, dirt, iron particles, and other solids that need to be removed. These kinds of particles are large and require a coarse sediment filter for removal. Water from a public water supply has already had the coarse particles removed, but it still has some degree of fine particles remaining. You can't see these fine particles, but they can neutralize the components of a water purifier. To eliminate this problem, a special sediment filter that can trap them is frequently used as the first stage of a drinking water purifier. This protects the other stages from getting clogged.

The following chart shows the relative sizes of common particles.

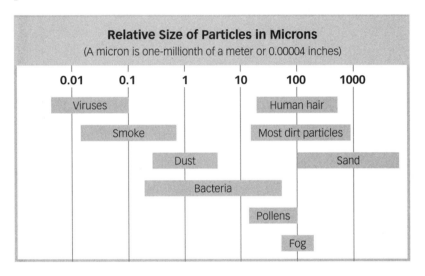

Note that the scale is exponential, not linear, so the chart is able to show a very wide range of size differences. For example, a sand particle is about a hundred thousand times as large as a virus. The length of the boxes in the chart shows the smallest to largest size of each type of particle; for example, pollen particles range from about 10 up to 100 microns. A micron is a very small unit of measure. For context, the diameter of a human hair is about 100 microns.

Sediment filters come in many different sieve sizes, from coarse to very fine. They are rated by the smallest size of particle they trap; for example, a 5-micron filter will trap all particles that are 5 microns or larger. Sediment filters for use on tap water from public water supplies (which has already been filtered) usually come in three sizes: 5, 10, and 20 microns. A 5-micron filter will provide complete protection for other stages of water purification. A 20-micron filter will

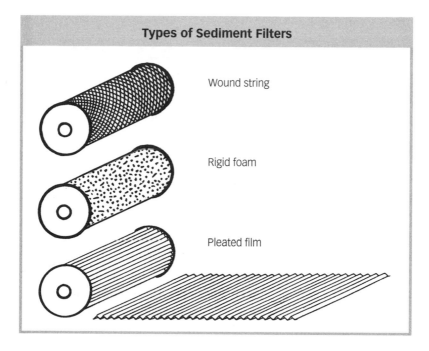

Types of Sediment Filters

Wound string

Rigid foam

Pleated film

provide slightly less protection but will last longer because it won't clog as quickly.

In a private system, if the water is visibly dirty a coarser filter will be required, typically 50 or 100 microns. This is followed by a finer, second-stage sediment filter that protects any additional water purification equipment in use.

For most home applications, there are three types of sediment filters available: wound string, rigid foam, and pleated film. The pleated film type is generally the most suitable for home use because it has the largest surface area and thus will last longer before becoming clogged with particles. Also, the thick, wet interior of the wound string and rigid foam types tends to support bacterial growth, while the continuous flushing action of water across the pleated film does not.

⬧ Carbon Filters

Almost all drinking water filters employ carbon as the primary filtration medium. Carbon filters are used to remove a wide variety of chemical pollutants from water. They are especially effective on organic chemicals, such as pesticides, herbicides, and industrial chemicals. They're also effective in removing radon, chlorine, and bad tastes and smells. When good-quality carbon filters are properly used, they remove 80 percent to 99 percent of the organic chemicals, radon, chlorine, and bad tastes and smells in water. However, they don't remove microorganisms or toxic minerals.

The type of carbon in water filters is *activated carbon*, which is made by heating wood, coconut husks, or coal in a special way so that millions of microscopic pores are formed. These pores attract and trap pollutants in water. There are two forms of activated carbon in wide use: granular and block. Carbon granules are about the size of coarse sand. Carbon block is finely powdered carbon that has been bound together into a rigid solid.

All other things being equal (such as size), a carbon block filter will remove higher percentages of pollutants and will last longer than a granular carbon filter. Because the granules in a granular carbon filter are free to move, when water passes through the filter it will try to find the shortest and easiest path, creating channels in the process. The result is that water flowing through the filter doesn't contact all of the carbon. Channeling can be minimized by proper design of the filter housing, but it can't be completely eliminated. On the other hand, a carbon block filter has the disadvantage of needing a sediment prefilter so that its dense pores don't become

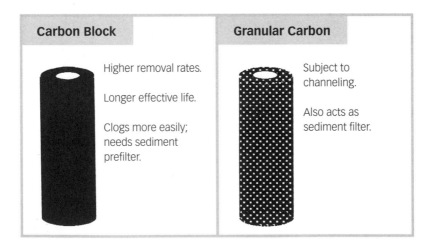

Carbon Block	Granular Carbon
Higher removal rates.	Subject to channeling.
Longer effective life.	Also acts as sediment filter.
Clogs more easily; needs sediment prefilter.	

clogged by dirt or mineral particles. A granular carbon filter, in contrast, acts as its own sediment filter.

Effective carbon filtration depends on several things. The most important are the amount of time the water is in contact with the carbon (the slower the flow, the more effective the filtration); the density of the carbon (carbon block is denser than granular carbon, and fine granular carbon is more effective than coarse granular carbon); the quantity of carbon (the larger the filter, the more effective the filtration); and the amount of water that has passed through the carbon (the fresher the carbon the better the filtration).

Like most filters, carbon filters accumulate pollutants within the filtering medium. This creates two potential problems. The first is that changes in water pressure can cause accumulated pollutants to break away and cause some unlucky person to ingest a large dose of pollution. The second drawback is that the inside of a carbon filter provides a supportive environment for the growth of certain kinds of bacteria. While these bacteria are not known to cause any disease symptoms, no one knows for sure how ingesting them affects human health.

Some manufacturers add a bacterial growth inhibitor to the carbon, usually a silver compound. Unfortunately, tests show that in actual use the silver does not do very much to reduce bacterial growth in most filters.

The potential problems of pollutant breakaway and bacterial growth can be minimized by replacing the carbon medium at regular, designated intervals and by not using the first flow out of the filter in the morning, after the water has been standing all night. When you first turn on the tap in the morning, you should allow the filtered water to flow for about thirty seconds before using it. This procedure should be repeated whenever the filter is unused for more than a few hours.

A carbon filter has a finite lifetime. A household of four people typically uses about 1½ gallons of drinking water per day, or 2½ gallons per day if filtered water is also used for cooking. This translates into approximately 500 to 1,000 gallons per year. Manufacturers of full-size carbon filters may claim their filters are effective for 2,000

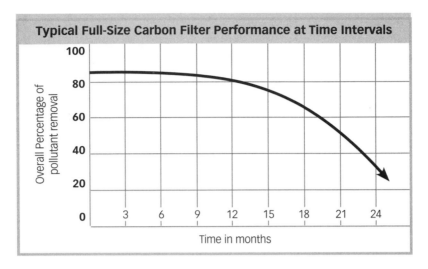

Typical Full-Size Carbon Filter Performance at Time Intervals

Overall Percentage of pollutant removal

Time in months

gallons or more, but I recommend changing carbon filters every six months. This ensures optimum performance and minimizes the chances of pollutant breakaway. Small carbon filters that attach to a faucet may have to be replaced every few months.

Some manufacturers include a valve in the filter so that water flow can be reversed. Called *backwashing*, this is supposed to regenerate the carbon so that it will perform as new. While carbon backwashing can remove much of the accumulated sediment within the filter, it generally is not effective in removing chemical pollutants from within the pores of the carbon. Because of this and because backwashing increases the potential for pollutant breakaway, I do not recommend backwashing carbon filters that are used for drinking water.

Sometimes carbon filters are installed in private water systems where the water has not been disinfected. This may result in a potentially harmful buildup of pathogens within the filter. In general, for a private water system, some method of disinfection should be used ahead of the filter.

When using a carbon filter, the main things to remember are:

- Avoid using the first flow; wait thirty seconds in the morning before collecting the water.

- When collecting the water, use a slow flow rate: ½ gallon per minute or less (the stream of water should always be narrower than a pencil).
- Replace the carbon at regular intervals as recommended by the manufacturer (or sooner). If the output water begins to smell or taste bad, discard the carbon immediately.

◉ Filters for Pathogens

Of the three basic classes of waterborne pathogens—viruses, bacteria, and parasitic cysts—bacteria and cysts can be removed by specially designed filters, but viruses cannot because they're too small.

In theory, carbon block filters should be able to remove bacteria and parasites from water because the micron rating of these filters—the smallest size of particle they trap—is smaller than most of the bacteria and all of the parasites found in water. But in practice the carbon is not uniform enough to ensure that *all* these microorganisms are trapped. Moreover, unless the seals within these filters have been specially designed to contain pathogens, they aren't 100 percent secure; some leakage may occur around the carbon medium. For these reasons, and because of the previously mentioned bacterial growth within the carbon, you shouldn't depend solely on a carbon filter for pathogen removal.

There is an important distinction between removing bacteria from water and simply stopping their growth within a filter. Many filters are labeled "bacteriostatic," meaning that bacteria don't multiply within the filter; this doesn't mean that the filter removes all bacteria from water. While many filter manufacturers claim that their product is bacteriostatic, these filters have generally not performed well when tested.

There are two types of filters that do remove essentially all bacteria and cysts from water: membrane filters and ceramic filters.

Both types come with positive seals, which means that there's no water leakage around the filter medium. Membrane filters use thin films whose pores are small enough to prevent bacteria, para-

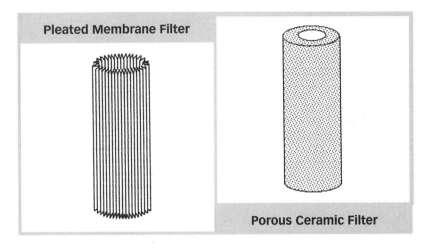

sites, and some viruses from passing through. While these filters are commonly used in labs, hospitals, and the like, they are usually expensive and not readily available to consumers. Ceramic filters use a porous, solid ceramic to trap pathogens. If the ceramic is of a high quality and consistency (same size pores throughout), its pores will trap bacteria, parasites, and some viruses. Ceramic filters tend to clog at the surface, but they can be periodically scrubbed many times over, and because of this they last for a long time.

A new type of UF (ultrafine) filter has recently become available. It's claimed to trap all cysts, bacteria, *and viruses*. Tests have confirmed this claim, but long-term reliability has not yet been established.

◆ Filters for Minerals and Metals

While some sediment filters and some carbon filters can remove solid particles of lead from water, these filters cannot remove lead or other toxic minerals that are dissolved in water. In recent years, several new kinds of filter media have been developed to remove dissolved lead and other toxic metals from water. Two of these have proved to be very effective: alumina filters and redox filters. Alumina is an aluminum compound that strongly attracts and traps metals that are dissolved in water. While alumina traps toxic metals, it doesn't remove chlorine, organic chemicals, or microorganisms.

Like carbon filters, the performance of an alumina filter decreases with use. When used for the specific purpose of removing lead from water, a small alumina filter installed in a kitchen lasts for about 2,000 gallons, or up to two years of average use.

The other addition to the types of filters available is the redox filter. The term *redox* is an abbreviation of *reduction/oxidation*, which is a chemical exchange process. In a redox filter, toxic metals in the water are exchanged for harmless zinc and copper. The zinc and copper in the filter also trap chlorine and smells caused by hydrogen sulfide, and these filters reduce (but don't totally eliminate) any pathogens in the water. One of the other advantages of a redox filter is that, unlike other filter media, it does not accumulate pollutants. Lead, chlorine, and other pollutants are converted to harmless zinc and copper compounds, which pass through the filter. Because of this, the zinc-copper medium within the filter has an extremely long life; recent tests have shown no reduction in effectiveness even after several years of use.

While other forms of water purification, such as distillation and reverse osmosis, can also remove lead and other toxic metals from water, until recently there has been no way to remove these pollutants with the simpler method of filtration. The alumina and redox filters are important and useful additions to drinking water treatment.

How Reverse Osmosis Units Work

Osmosis is the passage of molecules through the microscopic pores of a living or synthetic membrane. In normal osmosis, a difference in the concentration of molecules between one side of the membrane and the other, causes the molecules on the more concentrated side to pass through the membrane, equalizing the concentration on both sides.

Reverse osmosis (RO) water treatment employs a thin synthetic membrane with pores large enough to allow passage of water molecules but too small for larger molecules. Water pressure forces water

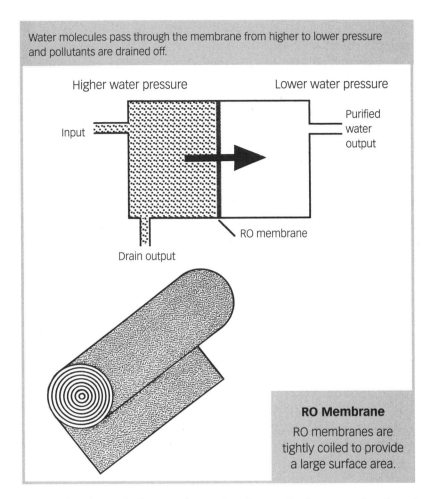

Water molecules pass through the membrane from higher to lower pressure and pollutants are drained off.

Higher water pressure

Lower water pressure

Input

Purified water output

RO membrane

Drain output

RO Membrane
RO membranes are tightly coiled to provide a large surface area.

molecules through the membrane but leaves the larger molecules of contaminants behind. The process is called *reverse* osmosis because, instead of equalizing the concentration of substances on both sides of the membrane, the water pressure creates pure water on one side and a concentrate of pollutants on the other. The pure water is channeled to the RO output, while water containing pollutants (as well as minerals) is channeled to a drain.

RO is a very slow process because the input water must pass through very small pores in the membrane. Although an RO membrane is tightly coiled within a cylinder to provide a large surface area, for home RO units it typically takes two to three hours for enough water to pass through the membrane to produce 1 gallon of water. Because the RO process is so slow, RO units use small storage tanks holding 2 to 5 gallons of water. When drinking water is drawn from the RO faucet, the water comes from the storage tank. The RO unit then slowly refills the tank.

Most RO purifiers waste water. Typically, they use between 3 and 10 gallons of tap water to produce 1 gallon of purified water; the remainder drains away. The amount of purified water produced in comparison with the total amount of water used is called the *recovery rate*, and most RO units have relatively poor recovery rates. However, a few RO designs are more efficient. In a *recirculating* RO unit, drain water is reused several times, increasing the recovery rate. There's also an in-line RO that uses the normal flow of water through household pipes in such a way that there isn't any water wastage.

The efficiency of RO units is also proportional to water pressure; the higher the pressure, the higher the ratio of cleaned water to wastewater. ROs installed in a systems with water pressure below 40 pounds per square inch (psi) require a special booster pump to increase the pressure.

One advantage of RO units over filters is that they remove a wider variety of pollutants. A good quality RO unit will remove 80 percent to 98 percent of most toxic minerals and organic chemicals from water. RO units cannot, however, remove radon or chlorine (while a carbon filter can). In theory, microorganisms, which are much larger than the molecule-size pores of an RO membrane, should all be rejected by the membrane. But in practice the pore size isn't uniform enough to ensure the removal of all microorganisms. Because of this, RO units, by themselves, cannot be used for disinfecting water.

There are several installation restrictions on RO units. They cannot be installed on a private system unless the water has been disinfected, because the RO membrane will accumulate a bacterial biofilm and become clogged. High levels of total dissolved solids (TDS) in water also adversely affect the membrane. Where tap water comes from a public water system and the water quality is known, an RO unit usually works well. But with private water systems, several water tests must be done before determining whether or not an RO unit can be used.

Unlike carbon filters, RO membranes don't accumulate pollutants—the pollutants are constantly being washed away. And since pollutants don't accumulate, there's no chance of pollutant breakaway as there is with carbon filters. However, the RO membrane itself degrades with use. With clean water from a public utility, a typical membrane lasts two to three years before it must be replaced. However, if there are bacteria in the water, if the water has a high TDS level, or if there are other adverse conditions, an RO membrane can fail prematurely, often after just a few months. Testers are available that can check the performance (and thus the condition) of an RO membrane in a few seconds, simply by pressing a button. These testers are inexpensive and should be installed with every RO unit.

There are two types of membranes for RO units: cellulose acetate (CA) and thin film composite (TFC). TFC membranes outperform CA membranes and last longer. They are also slightly more expensive, but their big disadvantage is that they cannot tolerate chlorinated water. With chlorinated water, an RO unit with a TFC membrane must be combined with an adequate prefilter to remove all chlorine.

While RO units provide an effective means of removing a wide variety of pollutants from water, they are almost never used alone. They invariably come as part of an integrated combination system in which filters are also employed. Combination systems are discussed in chapter 8.

How Distillers Work

Distillation is a simple, proven, dependable method of removing pollutants from water. Water is boiled, producing steam. The steam is cooled and condenses back into water. Any substances that do not evaporate are left behind in the boiling chamber. Distillation removes the widest variety of pollutants from water of any single purification method. But there is one type of pollutant that distillation does not remove well—the class of pollutants known as volatile organic chemicals. Because VOCs evaporate easily, it is more difficult for a distiller to separate them from the steam.

Like reverse osmosis, distillation is a very slow process. Depending on the size of the unit, it takes from two to six hours to make 1 gallon of distilled water. And like RO units, distillers store water in a tank or bottle for when it is needed.

Distillation also requires a lot of electrical power. Depending on the model of distiller and the price of electricity, it costs between 15¢ and 40¢(per gallon to make distilled water. (See chapter 10 for more cost per gallon comparisons.) They also generate a lot of heat, which may be a problem. If the distiller is located inside the house (as opposed to in the garage, for example), it will help warm the house in winter but will also add heat in summer.

One of the fundamental differences between distillation and all other types of water purification is that distillation is very reliable, producing consistently high-quality water with no decrease in performance over time. A ten-year-old distiller will produce the same quality of water as a new one. Also, the performance of distillers doesn't depend on manufacturing quality control. A filter, for example, may be advertised as performing up to a certain level, but the actual unit you buy may or may not perform that well, depending on the manufacturing standards. However, the simple fact that a distiller is more complicated than other types of purifiers and that it uses several electrical components means that it is more subject to potential malfunctioning.

There are two basic types of distillers made for home use: air-cooled batch distillers and water-cooled, continuous-flow distillers. Each type comes in one of three operating modes:

- Manual, air-cooled batch type
- Automatic, air-cooled batch type
- Automatic water-cooled type

Air-Cooled Batch Distiller

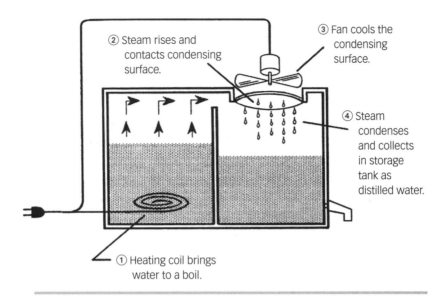

② Steam rises and contacts condensing surface.

③ Fan cools the condensing surface.

④ Steam condenses and collects in storage tank as distilled water.

① Heating coil brings water to a boil.

In batch distillers, the boiling chamber is filled with tap water and the distiller operates until all or most of the water is evaporated and collected in a storage tank. The distiller produces one batch at a time.

A heating coil heats the water until steam is produced. The steam is then directed to a condensing surface that is cooled by a fan. When the steam contacts the cool surface, it condenses back into water and drips into the storage tank. Very little water is wasted in this process. A batch distiller typically converts about 95 percent of tap water into distilled water, with 5 percent lost as vapor to the surrounding air.

Batch distillers can be manual or automatic. Manual distillers are filled by pouring water into the unit, and after the batch is completed, any remaining water is poured out manually as well. Automatic distillers are permanently connected to a cold water line and a drain line. They refill themselves after each batch is completed and drain away the residue water.

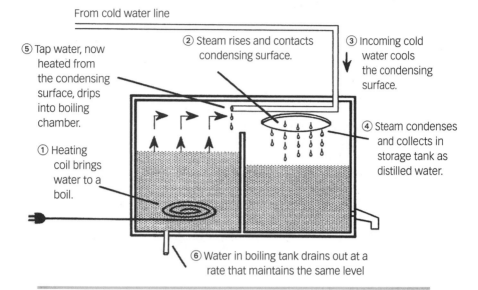

Water-Cooled Continuous-Flow Distiller

From cold water line

⑤ Tap water, now heated from the condensing surface, drips into boiling chamber.

② Steam rises and contacts condensing surface.

③ Incoming cold water cools the condensing surface.

① Heating coil brings water to a boil.

④ Steam condenses and collects in storage tank as distilled water.

⑥ Water in boiling tank drains out at a rate that maintains the same level

In a water-cooled distiller, the boiling chamber is always kept full by a constant trickle of incoming tap water. When steam is produced, it is directed to a condensing surface. After contacting the condensing surface, the steam reverts to water and drips into the storage tank. The condensing surface is kept cool by the incoming trickle of cold tap water. Because a fan isn't required to cool the condensing surface, water-cooled distillers are silent. They're also more energy efficient because the incoming tap water is heated by the condensing surface, so less energy is required to turn the water into steam; also, less heat is wasted since there isn't a fan transferring heat into the surrounding air.

Water-cooled distillers must be carefully adjusted so that the right amount of cold tap water flows into the unit. The incoming cold tap water controls the temperature of the condensing surface, so if the incoming flow is too high or too low, the condensing surface will be too cool or too hot for correct operation.

Now remember that distillers are good at removing all pollutants except VOCs. Because VOCs typically evaporate at lower temperatures than boiling water, if air vents are properly located, VOCs in the tap water will evaporate and be passed out through the air vents before the water boils. Although such air vents are a helpful feature, they cannot remove *all* volatile pollutants that may be in tap water. For this reason, most distillers are combined with carbon filters that *do* remove all VOCs.

How Ultraviolet Units Work

Ultraviolet (UV) radiation is higher in the electromagnetic spectrum than visible light. It is known to be effective in killing bacteria and other pathogens. The only types of pathogens that cannot be killed by ultraviolet are those with hard coverings, such as giardia cysts. Ultraviolet purifiers are made for the specific purpose of disinfecting water and are not effective in removing other pollutants.

Most materials, including common glass, do not transmit ultraviolet radiation efficiently. One of the best transmitters of UV radiation is quartz glass, so it is used for most UV lamps. In order for a UV unit to disinfect water effectively, three conditions must be met: (1) the UV lamp must produce radiation above a critical intensity; (2) the water must be subject to this radiation for a minimum period of time; and (3) both the input water and the UV unit itself must be clean. UV units employ many different designs to meet the first two conditions. The following illustration shows two simple designs: one with water flowing straight through a quartz cylinder past a long tubular UV lamp and the other with the water conduit wound in a spiral to increase the amount of time the water is exposed to the UV lamp. In the latter design, the spiral tubing must also be of quartz glass in order to transmit UV energy efficiently.

To meet the third condition, clean water, prefilters must always be used. If the water contains particles, some bacteria and other pathogens will be shielded from the UV radiation and will pass through the unit unscathed. Because water cleanliness is so critical to UV operation, all UV units for home use need to be designed to allow easy access for cleaning.

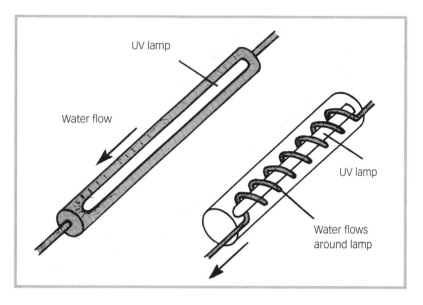

UV treatment adds nothing to water and takes nothing away. The advantage of this is that no toxic disinfectant remains in the water, as it does with chlorine disinfection. But that's also a disadvantage: If pathogens enter the water at any point *after* the UV treatment, there is no residual disinfectant to kill them.

Alternative Methods

In addition to the primary methods of purifying water at home— filtration, reverse osmosis, distillation, and ultraviolet radiation— other methods are sometimes used. Common alternative methods include deionization, aeration, ozonation, and air-to-water treatment. Deionization (DI) employs a tank of synthetic materials that attract and trap minerals as water passes through them. Like distillation, DI removes essentially all minerals from water, but in contrast to distillation, it doesn't remove other types of pollutants. DI is used when minerals must be completely removed from water and distillation is impractical. In general, DI is used in a whole-house water system to improve the smell or taste of the water, or when a particular mineral or metal needs to be removed.

Another method for removing certain pollutants is aeration. *Aeration* simply means exposing water to lots of air. This method has traditionally been used to remove bad smells from water, especially the rotten-egg odor of hydrogen sulfide. Aeration can also be used to remove radon and other volatile pollutants. Like DI, aeration is typically used in a whole-house water system to improve taste and smell.

There are two kinds of aerators: pressurized and atmospheric. Pressurized aerators employ air injectors that force air into a water line without losing water pressure. This type of aerator is not suitable for removing radon from water. An atmospheric aerator operates quite differently. It employs a nozzle that sprays a fine mist of water into an open tank. This allows volatile pollutants to evaporate into the surrounding air. This type of aerator is very effective at removing volatile pollutants. Its disadvantage is that when water is sprayed into the tank, water pressure is lost, so an additional pump is needed to repressurize the water.

Ozonation is sometimes used to disinfect drinking water in place of chlorination or ultraviolet treatment. Ozone is a toxic form of oxygen (O_3, rather than the O_2 of ordinary oxygen). It is a very powerful and effective disinfectant that works by chemically burning up pollutants (oxidation). In Europe and Asia, ozonation is commonly used in public water systems in place of chlorination. Ozonators can remove a wide variety of pollutants including pathogens, toxic organic chemicals and toxic minerals and metals. After being injected into water, ozone quickly reverts to normal oxygen and no residual ozone remains.

While there are some ozonators made for home use, the technology and manufacturing standards are critical because ozone itself reacts chemically to deteriorate many materials over time, so the components of ozonators must be chosen carefully.

Another fairly recent product development is the air-to-water purifier, which cleans air with special filters and then extracts the moisture from it. After moisture is captured and condensed, the water is passed through a series of conventional stages that may include filters, reverse osmosis, ultraviolet, and other treatments.

One advantage of air-to-water purifiers is that the moisture captured from air that has itself been cleaned is essentially free of pollutants even before it passes through the water treatment stages. Another advantage, with wide implications for the future, is that no water supply is needed to produce pure drinking water.

CHAPTER 8

Improving Performance by Combining Types of Purifiers

Because no single method of water purification can remove all potential pollutants, almost all water purifiers for home use combine more than one method. It is important to know the types of pollutants that various combinations will remove from your water; and remember, it is equally important to know what is likely to be in your water (see chapter 3) so you can choose the right purifier for your water profile.

There are two basic types of combination purifiers. The first type combines two or more treatment methods in a single, physically integrated unit. This kind of product is sold as one size fits all; there are no options. The second type consists of separate treatment sections that are interconnected. This type allows the consumer to select options for the separate treatment sections. In the case of filters this is made easier by the availability of standardized, interchangeable cartridges that perform different functions. Because these cartridges are easily interchangeable, a water purification system can be designed to work most effectively on your particular tap water.

Distiller Combinations

On their own, distillers remove almost all potential pollutants from water. The one exception is certain kinds of volatile organic chemicals (VOCs). Carefully designed distillers can even remove the majority of VOCs, but not all of them. But carbon filters are very effective at removing VOCs, and so the majority of distillers employ some type of carbon filter as a safety measure to capture any VOCs that the distiller misses. The carbon filter is either built into the distiller or is an attachment.

Some distiller manufacturers place the carbon filter ahead of the distiller (upstream) and some place it after (downstream). Placing it downstream is usually more convenient from a manufacturing standpoint. It also reduces the load on the filter because the distiller has already removed essentially all pollutants, so the filter will have a longer lifetime. But in this case, even though the distilled water entering the filter is free of pathogens, simply exposing the water outlet to air can contaminate the filter. (Note that any system that exposes a final-stage filter to air can become contaminated, although this is not a serious problem when compared to potentially harmful pathogens in the incoming water supply.) If the filter is placed upstream, it will tend to have a shorter life but the water coming out of the distiller will be totally free of pathogens. In my view, this approach is the better of the two.

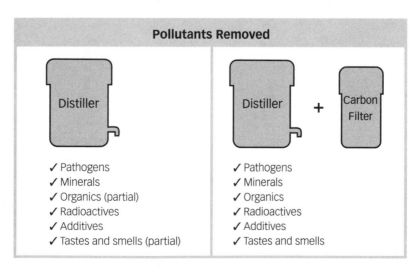

Pollutants Removed

Distiller	Distiller + Carbon Filter
✓ Pathogens	✓ Pathogens
✓ Minerals	✓ Minerals
✓ Organics (partial)	✓ Organics
✓ Radioactives	✓ Radioactives
✓ Additives	✓ Additives
✓ Tastes and smells (partial)	✓ Tastes and smells

Filter Combinations

As described in chapter 7, several types of filters are available for home use. The most commonly used are sediment, pathogen, redox, and carbon filters. Note the differences in performance for the combinations shown in the illustration. The first combination is a carbon filter with a sediment filter ahead of it to protect it from clogging. Together these do a good job of removing many pollutants, but they can't remove such things as pathogens and lead. Adding more filters improves the system's effectiveness. In the combination at the bottom, four different types of filters work together to remove almost all pollutants. While this system still can't remove certain substances such as nitrates and radioactive uranium, it is a very effective combination for most water conditions.

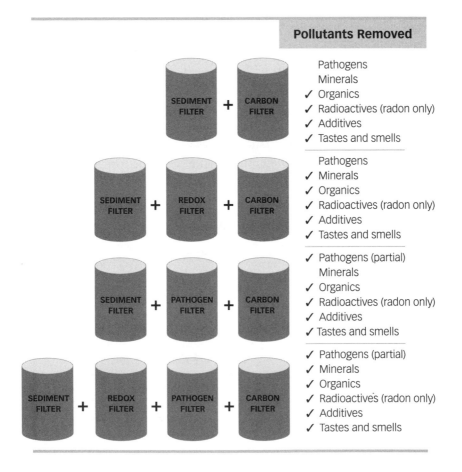

Pollutants Removed

SEDIMENT FILTER + CARBON FILTER
Pathogens
Minerals
✓ Organics
✓ Radioactives (radon only)
✓ Additives
✓ Tastes and smells

SEDIMENT FILTER + REDOX FILTER + CARBON FILTER
Pathogens
✓ Minerals
✓ Organics
✓ Radioactives (radon only)
✓ Additives
✓ Tastes and smells

SEDIMENT FILTER + PATHOGEN FILTER + CARBON FILTER
✓ Pathogens (partial)
Minerals
✓ Organics
✓ Radioactives (radon only)
✓ Additives
✓ Tastes and smells

SEDIMENT FILTER + REDOX FILTER + PATHOGEN FILTER + CARBON FILTER
✓ Pathogens (partial)
✓ Minerals
✓ Organics
✓ Radioactives (radon only)
✓ Additives
✓ Tastes and smells

Reverse Osmosis Combinations

Because of the delicacy of RO membranes, RO purifiers are always sold with a sediment filter to clean the upstream water. And because RO membranes do not remove all tastes, smells, and organic pollutants from water, these systems almost always come with a carbon filter. At the top of the RO illustration, notice that a basic RO system, with sediment and carbon filters, still does not remove all pollutants. Pathogens can get through, as well as some toxic minerals. Additional filters make RO systems more effective. Notice that the RO system at the bottom of the illustration is capable of removing almost all types of pollutants from water.

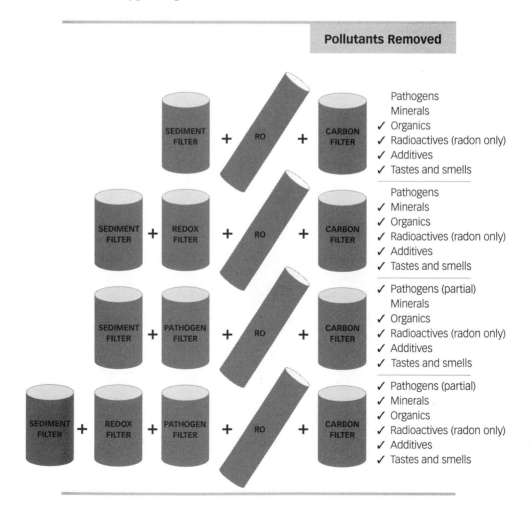

Pollutants Removed

SEDIMENT FILTER + RO + CARBON FILTER

Pathogens
Minerals
✓ Organics
✓ Radioactives (radon only)
✓ Additives
✓ Tastes and smells

SEDIMENT FILTER + REDOX FILTER + RO + CARBON FILTER

Pathogens
✓ Minerals
✓ Organics
✓ Radioactives (radon only)
✓ Additives
✓ Tastes and smells

SEDIMENT FILTER + PATHOGEN FILTER + RO + CARBON FILTER

✓ Pathogens (partial)
Minerals
✓ Organics
✓ Radioactives (radon only)
✓ Additives
✓ Tastes and smells

SEDIMENT FILTER + REDOX FILTER + PATHOGEN FILTER + RO + CARBON FILTER

✓ Pathogens (partial)
✓ Minerals
✓ Organics
✓ Radioactives (radon only)
✓ Additives
✓ Tastes and smells

Ultraviolet Combinations

As explained in chapter 7, UV purifiers are never used alone; they must be preceded by a sediment filter in order to be effective. In the illustration of UV combinations, notice that the top system removes only pathogens. Even within that category, it cannot kill giardia and some other parasitic cysts. The only time UV is used without additional filters is when its only purpose is to disinfect the water by killing bacteria and viruses, for example, when water is from an underground source and is very unlikely to contain cysts. If a carbon block filter is added to the system, it traps the cysts that pass through the UV unit unscathed, resulting in complete removal of pathogens. Note that the carbon filter used in this combination must be a carbon block filter, whose pores are small enough to trap cysts, and not a granular carbon filter. When a redox filter is added, toxic minerals are removed as well, and the system can

Pollutants Removed

SEDIMENT FILTER + UV

✓ Pathogens (partial)
Minerals
Organics
Radioactives
Additives
Tastes and smells

SEDIMENT FILTER + UV + CARBON BLOCK FILTER

✓ Pathogens
Minerals
✓ Organics
✓ Radioactives (radon only)
✓ Additives
✓ Tastes and smells

SEDIMENT FILTER + REDOX FILTER + UV + CARBON BLOCK FILTER

✓ Pathogens
✓ Minerals
✓ Organics
✓ Radioactives (radon only)
✓ Additives
✓ Tastes and smells

remove all pollutants except radioactive metals, such as uranium and radium.

Most tap water does not require extensive combinations of water purification methods in order to be reasonably safe, but some combination of methods is generally required. It's also important to know that some types of purifiers, employing a single method of purification, are sold with the claim that they will remove all harmful pollutants from water. This is simply not true.

CHAPTER 9

Evaluating Water Purifiers

Manufacturers' Claims

Although some states have legal standards for the performance of purifiers and the claims of manufacturers and sellers, there's still a lot of misinformation out there. One of the purposes of this book is to give you enough accurate information so that you will not be misled by false claims.

Most of the water purifiers sold in the United States are simple granular carbon filters because they are the cheapest and usually the easiest to install. Many filter manufacturers claim that these products can remove lead from tap water. This is misleading. Filters can trap lead in the form of particles, but most of the lead in water is dissolved lead, which carbon filters by themselves cannot remove.

For another example, take a look at the chart that shows misleading claims.

This chart supposedly compares the manufacturer's RO unit ("Brand X") with bottled water and carbon filters. In promoting its own RO system, this manufacturer has stretched the truth quite a bit. For example, the chart shows that Brand X RO removes bacteria and viruses. This isn't accurate. While ROs do remove some pathogens, they cannot remove them all and manufacturers cannot

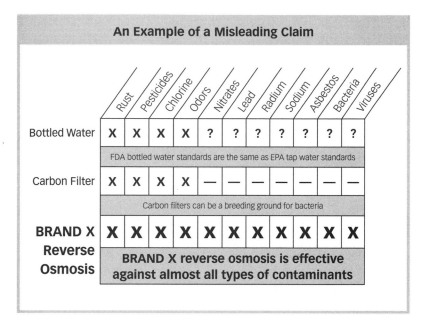

legally make this claim. ROs are also poor at removing nitrates from water, yet the chart claims that Brand X can remove them.

The comments within the chart are also misleading. In an attempt to discredit bottled water, the chart states that "FDA bottled water standards are the same as EPA tap water standards." While it is true that bottled water need only meet tap water standards under current federal law, it must also meet the requirements imposed by state laws and, for most bottlers, the standards of the International Bottled Water Association as well. The statement in the chart also ignores the fact that most bottlers use a multistage treatment process in addition to the treatment process the water utility already does. So in practical terms, the statement about FDA and EPA standards is meaningless.

The chart also states that "Carbon filters can be a breeding ground for bacteria." That is true. What is left unsaid is that the Brand X unit, which itself comes with a carbon filter, can also be a breeding ground for bacteria.

Another advertising trick is to use the word *bacteriostatic*. It's an impressive word, and it sounds like something that's very bad for bacteria. But as explained in chapter 7 it means only that bacteria will not *grow* within the unit; if there are bacteria in the incoming water, there will also be bacteria in the output water—bacteriostatic filters don't remove bacteria.

After surveying water purifier advertising for several years, I have come to the conclusion that much of it is incomplete, misleading, or just plain false. Sometimes manufacturers claim that their performance data is backed up by "laboratory tests." And, in fact, many manufacturers hire private labs to test their products. But you would need to be quite knowledgeable about test procedures to tell if the tests are meaningful or not. In general, you should not believe any performance claims or comparisons made by a water purifier manufacturer unless those claims have been certified by a recognized testing organization.

Product Testing Organizations

Whom can you rely on for accurate information about purifiers? It would seem that consumer-oriented magazines should be a good source of information. From time to time these magazines test water purifiers and report the results in their publications. I have monitored the test results in these publications over the years and have found them to be somewhat useful, but they also have some serious flaws. Typically, they test only a limited number of products, then generalize about the performance of all units of that type based on those few tested. Or they may test for a small number of pollutants and use those test results to generalize about overall product quality. That can be very misleading because the performance of products can vary widely with different pollutants. In general, I believe that the personnel involved in creating most magazine articles on water purifiers simply do not have the background needed to adequately report on this subject.

The American National Standards Institute (ANSI) is a longstanding and respected nonprofit organization that sets standards for U.S. products. Three private testing organizations are accredited by ANSI to test water purifiers to their standards. They are: the

Water Quality Association (WQA), Underwriters Laboratories (UL), and NSF International (formerly the National Sanitation Foundation). NSF is the most highly respected for testing; their staff is very knowledgeable, and the organization applies rigorous standards in all of their product tests. One good way to evaluate water purifiers is to look for the NSF certification seal in product advertising and literature. A listing of the types of NSF certifications appears in chapter 11.

The Water Purifier Performance Chart

The types of water purifiers available for sale vary greatly in their performance. For example, a small carbon filter that snaps onto a kitchen faucet has only limited effectiveness, while a larger carbon filter can do an impressive job of removing many pollutants. Given these differences, if performance ratings included the worst and best of products within any category of purifiers, the results would be meaningless—they would tell you very little about either the good products or the bad ones. Because of this, and because the purpose of this book is to guide you toward products that work well, the Water Purifier Performance Chart at the end of this chapter is based only on top-performing products.

Take a look at this chart right now. At first glance, it may seem awfully complicated, but don't be put off. You don't have to remember how all the purifier types work, and you don't have to remember what all the categories of pollutants are.

Some things pop out right away. You see from the chart that clear circles mean complete removal of a type of pollutant, all-black circles mean no removal, and partially black circles indicate some degree of removal—the more black, the poorer the removal.

A glance at the circles shows that the most effective water purifiers for removing *all* types of pollutants are the distiller plus filter combinations. Some of the other combination purifiers are also quite good.

At the other end of the spectrum, look at the fifth column from the right—a UV purifier by itself. UVs are made to kill pathogens and that's all they do, and even for pathogens, they do not kill parasites protected by cyst coverings. So unless bacteria and viruses are your sole concern, you would not use a UV purifier by itself. Now look at the column at the far right—a UV unit plus two types of filters, redox and carbon block. This combination performs quite well. Note that there are very few weak areas. And if you look at your Tap Water Profile (from chapter 3)—your record of what's likely or unlikely to be in your tap water and/or the water test report from your utility company—you may find it unlikely that it will contain any harmful amount of those pollutants not removed. Then if you look for a UV purifier, you don't need to know all the fine details, you simply need to find a UV purifier combined with redox and carbon block filters.

How Much Purifier Do You Need?

There are two ways of evaluating how much purifier you need, and you should consider both ways. The first way is comparing your Tap Water Profile (from chapter 3) to the performance of water purifiers. Remember, your Tap Water Profile is based on these factors:

- The test report from your public water utility (or on your own test results if you have a well or other private system)
- Whether your public utility is in a city or large town as opposed to a small town
- Whether your public utility gets its water from underground or surface water (or whether your private source is a well or surface water)
- The region where you live

Your Tap Water Profile helps you keep track of what you know about your tap water. That leaves what is unknown. The second way of looking at how much purifier you need is to realize that of the estimated eighty thousand chemicals in commercial use today, many of which have been found in water supplies, only a hundred or so are regularly tested for.

From this perspective, the safest course is to get a purifier that will remove any specific pollutants that you know are in (or have been in) your tap water, and that will also remove the most of all the many kinds of pollutants that may be in your water.

Water Purifier Performance Chart

○ = Complete Removal of Pollutant
● = No Removal of Pollutant

		GC Filter	CB Filter	REDOX + GC Filters	REDOX + CB Filters	B + C Filters	REDOX + B + GC Filters	Batch Distiller
Pathogens	Bacteria	●	●	◐	◐	○	○	○
	Viruses	●	●	◐	◐	●	◐	○
	Cysts	●	○	●	○	○	○	○
Minerals	Toxic metals	●	●	○	○	●	○	○
	Nitrates and other nonmetals	◐	◐	◐	◐	◐	◐	○
	Asbestos fibers	●	◐	●	◐	○	○	○
Organics	Volatiles	○	○	○	○	○	○	●
	Pesticides, PCBs, herbicides, and other nonvolatiles	○	○	○	○	○	○	○
Radioactives	Radon	○	○	○	○	○	○	○
	Uranium and radium, dissolved	●	●	○	○	●	○	○
	Uranium and radium particles	●	○	◐	○	○	○	○
Additives	Chlorine	○	○	○	○	○	○	○
	Fluoride	◐	◐	○	○	○	○	○
	Mineral alkalizers and flocculants	●	●	◐	◐	●	◐	○
	Organic additives	○	○	○	○	○	○	◐
Tastes and smells	Hydrogen sulfide and other volatiles	○	○	○	○	○	○	○
	Dissolved minerals	●	●	◐	◐	●	◐	○
	Mineral and organic particles	◐	◐	◐	○	○	○	○

Abbreviations used	B = bacteria	C = carbon filter
	CB = carbon block	GC = granular carbon
	RO = reverse osmosis	UV = ultraviolet

Evaluating Water Purifiers

	Water Flow Distiller	Batch Distiller + C Filter	Water Flow Dist. + C Filter	RO + GC Filters	RO + CB Filters	RO + B + C Filters	RO + REDOX + GC Filters	RO + REDOX + CB Filters	RO, REDOX, B + GC Filters	UV	UV + GC Filter	UV + CB Filter	UV + REDOX + GC Filiters	UV + REDOX + CB Filters
Bacteria	○	○	○	◑	◑	◑	◑	○	○	○	○	○	○	○
Viruses	○	○	○	●	●	●	●	●	○	○	○	○	○	○
Cysts	○	○	○	○	○	○	○	○	○	●	●	○	●	○
Toxic metals	○	○	○	○	○	○	○	○	○	●	◑	●	○	○
Nitrates and other nonmetals	○	○	○	◑	◑	◑	◑	◑	◑	●	●	●	●	◑
Asbestos fibers	○	○	○	○	○	○	○	○	○	●	●	◑	●	◑
Volatiles	●	○	○	○	○	○	○	○	○	●	○	○	○	○
Pesticides, PCBs, herbicides, and other nonvolatiles	○	○	○	○	○	○	○	○	○	●	○	○	○	○
Radon	○	○	○	○	○	○	○	○	○	●	○	○	○	○
Uranium and radium, dissolved	○	○	○	◑	○	○	○	○	○	●	●	●	○	○
Uranium and radium particles	○	○	○	○	○	○	○	○	○	●	◑	◑	○	○
Chlorine	○	○	○	○	○	○	○	○	○	●	◑	○	○	○
Fluoride	○	○	○	○	○	○	○	○	○	●	◑	◑	○	○
Mineral alkalizers and flocculants	○	○	○	○	◑	○	○	○	◑	●	●	◑	◑	○
Organic additives	◑	○	○	○	○	○	○	○	○	●	○	○	○	○
Hydrogen sulfide and other volatiles	○	○	○	○	○	○	○	○	○	●	○	○	○	○
Dissolved minerals	○	○	○	○	○	○	○	○	○	●	●	●	◑	◑
Mineral and organic particles	○	○	○	○	○	○	○	○	○	●	◑	○	○	○

Notes

1) All purifier combinations on this chart include a sediment filter as the first stage except for the distiller systems, which do not require one.
2) Performance ratings are based on top-performing products only—they do not include all available products.
3) In those systems where a granular carbon or carbon block filter work equally well, the general term "carbon filter" is used.

CHAPTER 10

Deciding What to Do about Your Water

Up to this point, we've looked at the potential problems with tap water, the merits and drawbacks of various kinds of bottled water, and the effectiveness of water purifiers. Now you're at the point where you can start deciding what to actually do about your tap water. When making your decision, in addition to considering performance, costs, and features of particular treatment systems, pay attention to convenience. There are a lot of water purifiers stored unused in closets and garages because their owners found them to be too much trouble. This chapter will help you anticipate how various types of purifiers will fit into your home and illuminate some of the advantages and drawbacks of each type of treatment system.

Cost Comparison of Bottled Water versus Purifiers

The Bottled Water versus Purifiers Chart compares the costs of various types of bottled water and using a purifier. An important point that bears repeating is that you can get high-quality drinking water both from bottled water and from home purifiers. By high-quality,

Bottled Water versus Purifiers Chart			
Type	**Cost**	**Type**	**Cost**
Bottled water from stores	$.35—$1.00 per gal.	**Bulk water from vending machines**	$.25–$.50 per gal. plus purchase of empty bottles
Bulk water, delivered	$1.40–$2.00 per gal. plus dispenser rental and bottle deposits	**Rent a purifier**	$30–$60* per month rental fee
Bulk water from water stores	$.35–$.50 per gal. plus purchase of empty bottles	**Buy a purifier**	Initial cost from $25–$1,500 plus $.06–$.62 per gal. depending on model

* For average family usage, cost per gallon is $.30–$.60 depending on rental charge.

I mean drinking water that is essentially free of most known pollutants and is at least as safe as the foods we ingest. However, some water purifiers—not all—can go beyond that and provide highly purified water from which essentially all potential pollutants have been removed.

If you don't want to bother with selecting, buying, and maintaining a water purifier and you opt for bottled water, you have four options: buy good-quality commercial bottled water; arrange for home delivery of bottled water; use your own bottles and buy purified water from a water store; or use your own bottles and buy purified water from a water vending machine. Bottled water is most convenient, of course, when it's delivered to your home so you don't have to lug heavy bottles into your car and home. But you pay a price for that convenience; the price per gallon for home-delivered water is higher than for most other bottled water.

If you've decided to use bottled water and you aren't interested in learning more about purifiers, you can skip the rest of this chapter and chapter 11.

Cost Comparison of Water Purifiers

The chart on page 99 compares costs of purchasing different treatment systems and also compares per gallon prices for treated water from each type of system. Bear in mind that purchase prices can vary greatly depending on the product you select. The purchase prices shown here don't include any installation costs. This is because within a given type of purifier, different models may or may not require installation; many purifiers just sit on a countertop or snap onto a faucet and are ready to use. If a purifier does require installation, it can often be done by anyone who is handy with basic tools. If a water professional installs it, expect a cost of $100 to $200 beyond the purchase price, depending on the complexity of the installation. Also, many dealers include installation in their prices.

All purifier combinations in the chart include a sediment filter, so that's not shown. Where "Carb. filter" is indicated, the filter may be either granular carbon or carbon block. The cost per gallon assumes that filters are replaced every six months, RO membranes every two years, and UV lamps every year. Here's a refresher on the abbreviations in the following chart: GC filter = granular carbon; CB filter = carbon block; Bact. = bacteria filter; RO = reverse osmosis; and UV = ultraviolet. (A redox filter removes minerals and metals.)

Does this cost comparison chart tell you anything meaningful? Well, the Cost per Gallon columns are informative because you can see that certain types of purifiers are more costly to operate than others over time. And while there's quite a range of prices shown for each purifier type, the chart tells you, for example, that some distillers are less expensive than some filters and that filters vary

Type of Water Purifier	Range of Prices	Cost per Gallon*	Cost per Gallon**
GC Filter Alone	$25–$200	6–12¢	8–25¢
CB Filter Alone	$75–$250	8–15¢	10–25¢
Redox + GC Filters	$125–$250	7–12¢	10–20¢
Redox + CB Filters	$125–$250	9–16¢	12–22¢
Bact. + Carb. Filters	$150–$250	13–22¢	16–25¢
Redox + Bact. + GC Filters	$200–$350	14–23¢	18–27¢
Distiller Alone	$120–$1,500	25–45¢	25–45¢
Distiller + Carb. Filter	$150–$1,500	27–47¢	29–49¢
RO + Carb. Filter	$120–$1,000	12–24¢	17–29¢
RO + Bact. + Carb. Filters	$450–$1,200	15–29¢	20–34¢
RO + Redox + Carb. Filters	$450–$1,100	14–27¢	19–32¢
RO +Redox + Bact. + GC Filters	$550–$1,300	16–29¢	21–34¢
UV Alone	$250–$800	4–7¢	6–9¢
UV + Carb. Filter	$300–$800	12–22¢	16–26¢
UV + Redox + Carb. Filters	$400–$800	13–23¢	18–28¢

PURCHASE AND OPERATION COST COMPARISON OF PURIFIERS

* If required replacements are performed by owner
** If required replacements are performed by a water professional

from an inexpensive $25 granular carbon pour-through pitcher to a more sophisticated $350 combination filter system.

Among the filters, the least expensive pour-through pitcher filters have the highest cost per gallon because the small canister filters need to be replaced frequently, typically every six to eight weeks. In general, the larger the filter, the greater the initial expense, but the lower the cost per gallon after purchase.

For distillers, the variation in cost per gallon is almost totally dependent on the cost of electricity in a given area. For RO systems, the variation in cost per gallon depends mostly on the cost of installing and replacing the RO membrane. For UV units, the cost per gallon is dependent on the replacement cost of the UV lamp and installation, if required.

The longer a filter cartridge or other replaceable component remains in operation, the lower the cost per gallon of making drinking water. Because of this, some manufacturers stretch their claims about the lifetimes of their products so that they can claim a low cost of operating the unit. However, as mentioned in chapter 7, carbon filters should be replaced every six months for optimum performance and safety, regardless of the manufacturer's recommendations, and redox filters should be replaced once a year. In the case of disinfecting filters, the manufacturer's instructions should be followed. As far as other purifier components are concerned, the average lifetime of an RO membrane is two years, and UV lamps last about one year.

The only purifier type that requires no periodic replacement of components is a distiller by itself with no filter. All other types employ cartridges, lamps, or other components that must be periodically replaced. Replacement of these components is easy for anyone who can use simple hand tools. On the cost comparison chart, the first column of cost-per-gallon prices (marked *) applies if you replace cartridges yourself or you can enlist someone to do it for you at no charge. If a dealer or other water professional replaces them, the cost for that service will increase the cost per gallon of making drinking water. The second column of cost-per-gallon prices (marked **) reflects the added charge for a professional to install all replacement cartridges.

Purifier Installation

Because there are so many different types of water purifiers, installing them runs the gamut. Pour-through pitchers require no installation whatsoever, whereas installing a whole-house system is a complex procedure for which you may want to hire a professional. Luckily, the majority of systems designed for home use aren't difficult to install, and most people who are comfortable using hand tools can do the job easily.

Whole-house systems are referred to as *point-of-entry* (POE) systems, meaning that they treat the home's entire water supply before it's distributed throughout the house. The most common POE systems are water softeners. Others are used in certain circumstances; for example, for removing radon, chlorine, or other volatile chemicals from the water. They're also sometimes used to improve the water's aesthetic qualities by removing excessive minerals and bad tastes and smells. POE systems typically consist of relatively large tanks and, often, controls for backwashing of the tank contents. But in most cases, it's only necessary to treat water for drinking, in which case you can install a point-of-use (POU) system, usually in the kitchen. Several small filters, and a few ROs, simply snap onto a kitchen sink faucet and require no other installation. For the rest, the most common types of POU installations are described below.

◆ Purifier Installation Type 1

The simplest type of installation is shown in illustration 1. This is a countertop RO unit or filter with a tube from the faucet. The water connection at the faucet usually replaces the faucet aerator. When a button or valve on the faucet is pushed, water flows to the purifier and out the spigot on the purifier. If an RO is used, a drain line into the sink is also provided.

◆ Purifier Installation Type 2

In illustration 2 (see page 102), a countertop filter or RO is connected to the faucet by a double tube. When the diverter is moved,

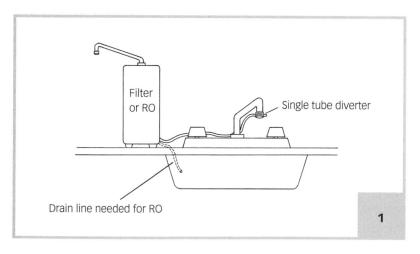

Filter or RO

Single tube diverter

Drain line needed for RO

1

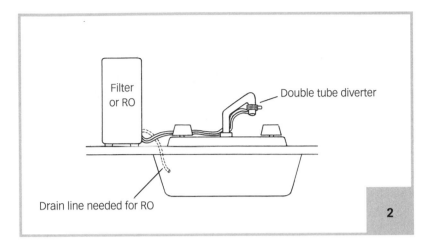

Filter or RO

Double tube diverter

Drain line needed for RO

2

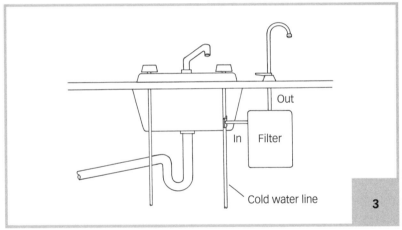

Out

In | Filter

Cold water line

3

tap water flows from the faucet to the purifier in one tube, and treated water returns through the other tube and flows out of the sink faucet.

● Purifier Installation Type 3

In illustration 3, a filter is installed under the sink. It receives water from the under-sink cold water line, and the treated water is connected to a separate countertop faucet. Under-sink installations are convenient because the purifier isn't on the countertop, where it can get in the way, and there are no awkward tubes connected to the sink faucet.

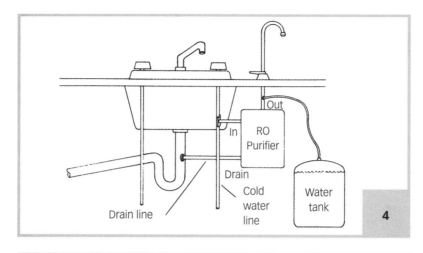

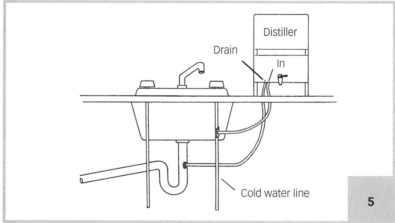

● Purifier Installation Type 4

In illustration 4, an RO is installed under the sink. It receives water from the under-sink cold water line, RO wastewater is connected to the under-sink drain line, and the output of the RO unit is connected to a storage tank. When the RO produces water, it fills the storage tank and enlarges an expandable bladder within the tank. The enlarged bladder exerts pressure on the water, which goes up to a separate countertop faucet so that water will flow to the faucet as long as there is some water in the tank.

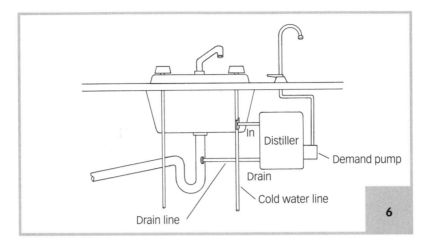

Demand pump

Distiller

In

Drain

Cold water line

Drain line

6

◆ Purifier Installation Type 5

In illustration 5, an automatic distiller is installed on the counter-top. It is connected to the under-sink cold water line and to the under-sink drain. An internal storage tank in the distiller stores water and has its own spigot. An alternative is to install an automatic distiller in a remote location, such as a utility room or garage. In this case, provision must be made for access to a cold water line and a drain line.

◆ Purifier Installation Type 6

In illustration 6, an automatic distiller with its own storage tank is installed under the sink. It is connected to the under-sink cold water line, to the under-sink drain, and to a separate countertop faucet. When the lever on the faucet is pushed, it activates a small pump (called a *demand pump* because it operates only when the faucet lever is pressed), delivering distilled water to the faucet. In this installation, the distiller can also be in a remote location where a cold water line and a drain are available.

Living with a Water Purifier

Each type of water purifier has its own characteristics, and you may find some of them easier or harder to live with. Here are some helpful things to know about before you buy a purifier.

◆ Distiller Characteristics

If you're planning on installing a batch-type distiller in a permanent location, be aware that they give off as much heat as a small portable electric heater and also add to the air's humidity. Portable countertop distillers can be filled and then moved anywhere. Also, batch-type distillers are cooled by a fan, so they're noisy. Most distillers for home use require three to five hours to make 1 gallon of distilled water, which means that the fan will remain on for long periods. For permanent installation, automatic distillers are typically placed in a utility room, garage, or other out-of-the-way location.

Batch distillers are water-efficient, whereas continuous-flow distillers waste several times more water than they make. You should consider this if you live in an area where water is scarce. On the other hand, continuous-flow distillers are essentially silent and give off far less heat than batch distillers.

Another characteristic of distillers is that the treated water comes out hot. If you're out of drinking water and waiting for the next batch with bated breath, you'll have to cool the water down. And since a distiller requires electricity, if the power goes out, you can't use it to purify water.

Because distillers have more components than other purifiers, some are prone to breakdowns. So it's important that you purchase a distiller with a solid record of dependability.

Distillers require almost no maintenance when the household water is soft. With hard water, scale builds up and hardens in the boiling chamber. Because of this, easy access to the boiling chamber is one of the most important features of a distiller.

For all their inconveniences, distillers in combination with carbon filters produce the most consistent and reliably safe water of any purification process. If the distiller is working, it is always working 100 percent. With periodic replacement of the carbon filter, the water produced by a distiller will be as pure after ten or twenty years of service as it was when the distiller was new. And other than the cost of replacing carbon filters, the only expense involved in operating a distiller is the cost of electricity.

● Reverse Osmosis Characteristics

Most reverse osmosis units fall into two basic types: simple, relatively inexpensive snap-on faucet units, and larger, multistage units that can be installed on a countertop or under the kitchen sink.

Like distillers, RO units make water very slowly. If you need several gallons of drinking water at one time, this will empty your storage tank, and you will have to wait several hours for the RO to produce more water.

The small faucet-mount units consist of a simple sediment filter and an RO membrane; some units contain a small carbon filter as well. While the faucet units do not remove as many pollutants as the more sophisticated multistage units, they have the advantage that they can be stored in a refrigerator when not in use. This is an advantage because all RO membranes develop some degree of biofilm—a thin layer of microorganisms—and refrigerator temperatures inhibit this biofilm growth.

Almost all multistage RO systems come with two or more filters (see the Water Purifier Performance Chart in chapter 9 for details). But even with filters, RO membranes can degrade and/or foul easily if the tap water exceeds certain limitations (temperature, dirtiness, acidity, and so forth). If you plan to go with a multistage RO unit, you should purchase it from a dealer who knows local water conditions and is experienced with RO systems. Make sure the dealer knows what is in *your* water and your water pressure before installing an RO system. In general, ROs should not be used on private water systems unless the water is first treated to remove or kill pathogens.

ROs waste water—some more than others. Some ROs keep making and wasting water even when the storage tank is full. Several states now have laws requiring RO systems to have automatic shutoff valves that stop the flow of water when the tank is full. If you live in an area where water is scarce, check out the efficiency, or recovery rate, of the RO you are interested in. Some ROs are very inefficient, with a recovery rate of 1 to 8 or more, meaning that they waste eight times as much water as they make. Highly efficient units have recovery rates as low as 1 to 2, and a few models don't waste water at all.

RO membranes are delicate and subject to possible degradation under certain conditions. If you buy a multistage RO system, I recommend that it includes an output monitoring device. These simple meters that measure how efficiently the RO membrane is working are inexpensive and easy to install. Many ROs come with a built-in monitoring device. If the model you're considering doesn't include one, ask your dealer to add it to your system.

ROs operate on water pressure, and the higher the pressure, the more thorough the rejection of pollutants and the more efficient the production of pure water. Most ROs require at least 40 psi (pounds per square inch) of water pressure for effective operation. The pressure in a typical home is above this, usually 50 to 70 psi. Low pressure is typically present in older homes where the water pipes are corroded, or with private water systems where the water must be pumped to the house. The dealer should check your water pressure before installing an RO system. If the pressure is too low, a miniature booster pump can be added to the RO unit.

There are two types of membranes available for ROs. The first is called a CA (cellulose acetate) membrane. The second type is called a TFC (thin film composite) membrane. The CA type is less expensive, works well on most chlorinated tap water and does a good job of removing pollutants. TFC membranes outperform CA membranes, but they're more expensive (adding $25 to $50 to the cost of an RO system); plus, they can't tolerate chlorinated water, so a carbon or redox prefilter is necessary to remove chlorine before it reaches the RO membrane. Where a CA membrane might remove 90 percent to 95 percent of most pollutants, a TFC membrane will remove about 97 percent. TFC membranes also last longer. Before you buy an RO system, ask the dealer which membrane type is best for your conditions.

RO units don't need electricity, making them much more energy efficient than distillers. While a distiller plus filter system achieves the most thorough removal of all pollutants from water, a good-quality RO system that includes filters is silent, doesn't use electricity or generate heat and humidity, and comes close to a distiller system in effectiveness.

◆ Filter Characteristics

Filters by themselves cannot remove the total spectrum of possible pollutants from water. But good-quality filters can often remove the most harmful pollutants from water, and filtered water is still better than untreated tap water. Filters are relatively inexpensive to buy and inexpensive to operate. In addition, they're more convenient than either ROs or distillers because they produce water—as much as you need—immediately. This also means there's no need for storage tanks.

Portable, pour-through filters are the most popular home water treatment device in the United States. This is because they are cheap, they require no installation, and they produce water that tastes pretty good. These portable filters consist of a small canister filled with granular carbon and usually some kind of thin membrane to prevent the carbon from getting into the output water. Their downside is that their small size limits their effectiveness (remember, size and the amount of contact time within the filter are key to effectiveness) and the fact that carbon filters, by themselves, can only remove limited kinds of pollutants. However, these filters do have an upside: Unlike almost all installed filters, they can easily be refrigerated, which retards growth of microorganisms within the filter. In my view, this is a real advantage. Before each use, I recommend that the canister be flushed for at least ten seconds before filtering another batch of drinking water.

The second most popular water treatment devices are the small, self-contained filters that fasten onto the end of a kitchen faucet. Most of these miniature filters have the same innards as the pour-through filters and thus have the same limitations. Faucet filters are typically attached with an easy snap-on connection. Although manufacturers don't recommend or even mention it, most faucet filters can be removed and stored in the refrigerator when not in use. For the sake of convenience, filter a batch of water before doing so, then store both the water and the filter in the refrigerator. After storing a faucet filter in the fridge, remember to flush it for at least ten seconds before use.

Also, some snap-on faucet filters, slightly larger in size, have recently come on the market with additional filter stages that improve performance.

Overall, miniature pour-through and faucet filters do improve tap water. They improve the taste and can remove bad smells if they're not strong. They also remove most of the chlorine and some, but not all, other pollutants. They are inexpensive, too, but since the filter canisters need to be replaced every six to nine weeks, the cost per gallon of treated water is higher.

Other filters come in all shapes, sizes, and prices. Most of the larger, better-quality filters employ several different types of filter media, so they perform much better than the miniature filters. But the majority of these larger filters use proprietary housings and cartridges, so you are locked into buying all replacement cartridges from that manufacturer—and as you might guess, the replacements are

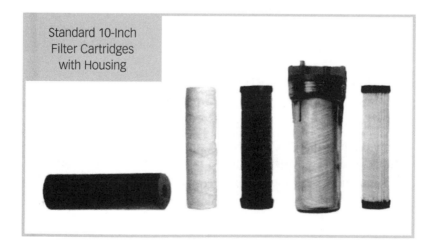

Standard 10-Inch Filter Cartridges with Housing

not cheap. Another disadvantage of these proprietary filters is that they can't be customized to accommodate all water conditions.

There is an alternative, and it has many advantages over the proprietary-type filters. Hundreds of manufacturers make standard, 10-inch filter cartridges that fit into standard filter housings. (Note that some manufacturers use a 9¾-inch designation instead of 10-inch; both of these refer to the same standard cartridges. Also, some standard cartridges are called *candle style*. These cartridges, thinner than most standard cartridges, are interchangeable with other standard cartridges and fit into the same housings.)

Standard 10-inch cartridges are of adequate size, good quality, and come in several varieties so that a purifier system can be customized for particular water conditions. They are relatively inexpensive, are available everywhere, and are all interchangeable. That means you are not tied to any particular manufacturer for replacements. The cartridge housings are also standardized, inexpensive, and can be stacked together to create anything from a simple system to a multistage, highly effective one.

An optional feature to consider when buying a filter system is a flow meter, which counts the gallons of water produced. When the filter has processed a specified number of gallons, the flow meter either turns off the water supply or alerts you, for example with a red light. These are helpful in reminding you to replace filter cartridges and add only a modest amount to the cost of a filter system. If you are reasonably careful about scheduling cartridge replacements, you may not need this option.

● Ultraviolet Characteristics

Ultraviolet purifiers have a single purpose—to kill bacteria and viruses. UV units have traditionally been used on private water systems that may contain these microorganisms. For this purpose, UVs compete effectively with chlorinators and iodinators that add disinfecting chlorine or iodine to the water. They also compete with ozonators—devices that generate ozone gas and insert it into the water as a disinfectant. For disinfecting water, a UV has a distinct advantage in that it adds no harsh chemicals to the water that may, in themselves, be toxic.

Although there are a few portable countertop UV units, almost all of the models available are for under-sink or remote installation. All UVs run on electricity, so if the power is out, they can't be used to disinfect water.

UV lamps have a limited lifetime—usually one year of constant operation. Some units have a feature that turns the lamp on only when water is being drawn through the unit. This feature extends the life of the lamp, but eventually all UV lamps have to be replaced. So an important feature of a UV is whether or not the lamp is easy to replace. Also, UV lamps (or the clear quartz covers that protect them) get dirty and must be cleaned occasionally. Some UV units have built-in wiper arms, allowing the lamp to be cleaned from the outside of the unit without taking anything apart; others must be partially disassembled for cleaning. An important feature of a UV unit is an alarm or shutoff device that kicks in when the intensity of the UV energy falls below a critical level for any reason.

A UV unit is a specialized device. It doesn't have the ability to remove the wide variety of water pollutants that other types of purifiers do. If you check the Water Purifier Performance Chart in chapter 9, you will notice that most UV systems show a lot of completely black circles, indicating no removal of those kinds of pollutants. In order to remove a wide variety of pollutants, you need to combine a UV with several filters, as in the column on the far right of the chart.

When would you choose a UV system? If you are on a private water system that is unchlorinated, UV is one of the surest and safest ways to disinfect your drinking water. If the source for your private system is surface water, you will need a UV system with a carbon block filter or a bacteria filter to trap any possible cysts in the water, which UV does not kill.

A UV system with a carbon block or bacteria filter is also a good solution when your tap water comes from a public water supply that has been disinfected, but there have still been outbreaks of disease from the water. In this case, such a UV system would be an effective means of killing waterborne bacteria and viruses and trapping cysts. As outbreaks of illness within communities are increasingly traced to waterborne microorganisms, the use of a UV system becomes increasingly attractive.

◆ Taste

The presence of chlorine, of course, affects the taste of water, and if you are used to drinking chlorinated water you will probably be pleasantly surprised at how much better it tastes with the chlorine removed. Beyond that, different types of purifiers affect the taste of water in different ways. Distilled water removes all minerals, producing water that some people describe as tasting flat, while others don't detect much difference. Carbon filters, on the other hand, retain minerals. But in addition, carbon filtration somehow (and it is not known exactly how) seems to improve the taste of water, and the output water from this process is often described as tasting fresh and bright. Like distilled, RO membranes also produce a flatter-tasting water, but carbon filters are almost always included downstream of the RO, so this usually isn't a factor. Ultraviolet units, by themselves, don't change the taste of water. If taste might be a factor in making your selection, you can try bottled distilled water and bottled drinking water, and that will give you an idea of the difference.

Purifiers: Should You Buy or Rent?

A rental unit can be a viable option for improving your drinking water. It saves you a large, up-front cash outlay, and if you are interested in eventually buying a multistage purifier, renting one for a while allows you to live with it before deciding to purchase your own.

Renting a water purifier is not an option in all areas. If you are interested in renting a water purifier, call the dealers in your area to find out if they have rental programs. Distillers, multistage filters, UVs and ROs may be available for rental. Monthly rental fees average about $25 to $40. This should include service calls and the periodic replacement of cartridges, RO membranes, or UV lamps. There may be an installation charge, or you may be required to sign a rental agreement for a given amount of time.

If you rent an RO system, it is important to monitor the performance of the RO membrane; you usually can't tell what condition it's in by the taste or appearance of the water. If the RO unit does not have a built-in monitoring meter, a service person should check the RO membrane at least every six months.

If you are considering renting a distiller, you should probably do so only if you can provide softened water to the distiller. Hard water causes rapid buildup of scale in a distiller, and frequent (sometimes difficult) cleaning is required.

CHAPTER 11

Selecting a Water Purifier

Chapters 7 through 10 described the general characteristics of different types of purifiers—how they work, what pollutants they remove, general costs, the advantages and disadvantages of different systems—to help you determine which type of purifier will work best for you and your tap water. This chapter gives you specific information on recommended manufacturers and models that will help you decide which particular purifier to purchase.

Cautions and Recommendations

Before the advent of the Internet, it wasn't always easy for consumers to obtain information on water purifiers from manufacturers and distributors, and the only way to buy purifiers was through local dealers who offered only a very limited number of products. In contrast, there are now thousands of websites that not only provide a glut of information (not always accurate), but also sell directly to the consumer. This chapter is designed to be supplemented by the Internet. For most products, simply type in the brand name of a purifier or generic descriptive terms (for example, "countertop reverse osmosis") in a search engine and you'll find a multitude of websites that have detailed information and pictures, plus more models and options than can possibly be mentioned in this book.

◆ Online Sales and Sales at a Distance

While the Internet offers you an abundance of information, there are advantages that come from dealing locally.

If you purchase a water purifier locally, you'll have the opportunity to talk to a salesperson who can help you choose a model appropriate to your situation. They will also have a better grasp of any considerations specific to local water supplies. If you purchase online, this type of customer service isn't available. The disadvantage of dealing locally is that the selection of purifiers will be much more limited.

Filters of every type can safely be ordered online. The only restriction to this is if your water supply is unchlorinated (from a private well or spring) and you are not certain if the water is free of pathogens. In this instance, you need to employ a system that totally disinfects your water, not one that just reduces the pathogen level. As a quick reminder, only two types of purifiers will disinfect water with certainty: distillers and ultraviolet systems in combination with cyst-trapping filters.

Since there are relatively few distiller dealers, you may not be able to purchase a distiller locally. However, distillers can safely be ordered online or from a distant dealer. Just remember that distillers are more complicated devices than other types of water purifiers and that some distillers are highly dependable and others are not. So when choosing a distiller, be careful what you choose and pay attention to warranties and company service policies. The companies who manufacture the distillers listed in this chapter are all recommended for their products, warranties, and service.

Small, faucet-mounted ROs are generally suitable for use on any water supplied by a utility company and can be ordered online. If the RO membrane in this type of purifier should deteriorate, it's relatively inexpensive to replace. If you are considering a larger RO combination system, it's much better to order it from a local dealer who knows your water conditions. If you don't purchase it locally, you *must* make the effort to ensure your tap water is compatible with the RO system you buy.

If your water is supplied by a utility company, you can safely order a UV unit online. If your water supply is from a private well or spring, a UV purifier must be preceded by a sediment filter.

And lastly, there are thousands of online suppliers of standard 10-inch filter cartridges, which will enable you to customize your water purifier system if you so choose.

◆ Product Certifications

As explained in chapter 9, there are several accredited, independent organizations that test water purifiers. If you're considering a purifier other than the models recommended in this book, a good way to verify the effectiveness of a purifier is to look for an NSF/ANSI (National Sanitation Foundation International/American National Standards Institute) seal of approval for the type of purifier you are interested in:

- Look for NSF/ANSI 42 certification for filters for removal of chlorine, tastes, and smells.
- Look for NSF/ANSI 53 certification for filters for removal of other pollutants.
- Look for NSF/ANSI 55 certification for all ultraviolet units.
- Look for NSF/ANSI 58 certification for all reverse osmosis units.
- Look for NSF/ANSI 62 certification for all distillers.
- Look for NSF/ANSI 177 certification for all shower filters.
- Look for NSF/ANSI P231 certification for all specialized disinfecting filters.

The above ANSI standards are also certified by the WQA (Water Quality Association) and UL (Underwriter Laboratories), so the approval of these organizations is also satisfactory.

The above certifications are expensive to obtain and many manufacturers of quality purifiers and components do not have them, so lack of certification doesn't necessarily indicate an inferior product.

Sometimes a manufacturer will claim a certification for the components of a purifier. In the case of filters, for example, the filter media may have been certified by the NSF and the parts ISO certified (International Standards Organization), but the performance of the overall unit has not been certified. Even if a purifier may employ quality materials, sometimes the manufacturing process creates imperfect seals, allowing untreated water to leak into the treated output water. And some manufacturers use glues and nonfood-grade materials that can actually leach toxic pollutants into the treated water. So if you see a certification, make sure the certification is for the entire purifier and not just for some of its components.

In general, if you are interested in a purifier that is neither certified nor included in the recommendations in this book, at least make sure that the manufacturer is a long-standing, reputable one. That, in itself, provides a good deal of assurance.

◆ Ratings

If we could afford the very best water purifier and the most complete safety of drinking water was the sole criterion in choosing one, it would be easy to recommend just a few purifiers. But cost is a factor, as are convenience, maintenance, and individual preferences. So the recommendations that follow are for the most effective and cost-effective purifiers in each of the various categories. For example, the rating for a small, inexpensive unit will be lower than for a larger, sophisticated one.

The characteristics and relative convenience of different types of purifiers are discussed in detail in chapter 10. The rating at the end of each of the descriptions below is based on the unit's effectiveness in eliminating all potential pollutants. There are four levels of recommendation:

- **Acceptable**—Improves tap water
- **Good**—Eliminates the majority of potential pollutants
- **Very Good**—Eliminates almost all potential pollutants
- **Excellent**—Among the most effective purifiers

Filters

� Pour-Through Pitcher Filters

There are several pour-through pitcher filters on the market for home use, with two or three of the most popular ones available in most hardware stores, building supply stores, drug stores, discount stores, and the like. All of these filters use proprietary cartridges.

CRYSTAL QUEST PITCHER. Has the highest degree of filtration for a pitcher filter, with microfilters, a carbon filter, and an ion-exchange filter for reducing lead.

Price range: $25–$35, depending on model

Replacement cartridges: $10–$13, depending on quantity purchased

Cartridge life: 6 to 12 months claimed (6 months maximum recommended)

Rating: Acceptable

PUR CR-800, CR-900, AND DS-1800. All have a carbon filter and a microfilter for trapping cysts.

NSF 42 and 53 certified

Price range: $26–$50, depending on model

Replacement cartridges: $8–$14 each, depending on quantity purchased

Cartridge life: 2 months with average use

Rating: Acceptable

PUR CR-1500R AND CR-1510. Both have a carbon filter and an ion-exchange filter for removing lead.

NSF 42 and 53 certified

Price range: $19–$27, depending on model

Replacement cartridges: $13–$18 each, depending on quantity purchased

Typical Pour-Through Pitcher Filter

Cartridge life: 2 months with average use

Rating: Acceptable

BRITA, ALL PITCHER MODELS. All have a carbon filter and an ion-exchage filter for reducing lead.

NSF 42 and 53 certified

Price range: $10–$35, depending on model

Replacement cartridges: $6–$8 each, depending on quantity purchased

Cartridge life: 2 months with average use

Rating: Acceptable

● Faucet Filters

There are dozens of different faucet filters available that are minimally effective and just a few that are slightly larger and more effective. All faucet filters use proprietary cartridges. The more effective ones are listed below.

CRYSTAL QUEST W2. Has a carbon filter, a redox filter for lead reduction, a cyst-blocking filter, and an ion-exchange filter for further lead reduction.

Price range: $25–$40

Typical Faucet Filter

Replacement cartridges: $10–$13 each, depending on quantity purchased

Cartridge life: 6 to 12 months (6 months maximum recommended)

Rating: Acceptable

PUR FM-9100, FM-9400, FM-9500, and FM-9800. All have a carbon filter, a redox filter for lead reduction, and a cyst-blocking filter.

NSF 42 and 53 certified

Price range: $48–$60

Replacement cartridges: $9–$13 each, depending on quantity purchased

Cartridge life: 3 months with average use

Rating: Acceptable

CULLIGAN FM-5. Has a carbon filter, reduces lead, and blocks cysts.

NSF 42 and 53 certified

Price range: $45–$56

Replacement cartridges: $20–$25 each, depending on quantity purchased

Cartridge life: 3 months with average use

Rating: Acceptable

GE GXFM03C and GXFM04FBL. Both have a carbon filter, reduce lead, and block cysts.

NSF 42 and 53 certified

Price range: $22–$30

Replacement cartridges: $14–$18

Cartridge life: 3 months with average use

Rating: Acceptable

◆ Countertop Filters with Proprietary Cartridges

There are dozens of countertop water filters available that use proprietary cartridges and either attach through tubes to a kitchen faucet or have their own faucet. Many of them are quality purifiers made to high manufacturing standards. But for most of them a disadvantage is the high cost of their proprietary replacement cartridges. Ones that stand out for being effective, high quality, and a good value are listed below.

AQUASANA AQ-4000. Has a first-stage sediment/carbon filter and a second-stage redox/cyst filter. One of the best small countertop filters. Attaches to kitchen sink faucet.

NSF 42 and 53 certified

Price range: About $120

Proprietary cartridges: $25–$30

Cartridge life: Approximately 6 months

Rating: Good

CRYSTAL QUEST PLUS. A larger countertop unit that is sealed and has no replacement cartridge. It has its own faucet. It employs two cyst filters, two types of redox media, ion-exchange media, and granular carbon media. Its larger volume allows more effective filtration than most countertop units. It is also backwashable.

Filter media NSF certified

Price range: $90–$130

Aquasana Countertop Filter

No cartridge replacements

Filter life: 2 to 3 years claimed. While most filters should be discarded after 6 months in spite of manufacturers' claims, if this unit is used on treated municipal water and is back-washed every few months, the recommendation is extended to 1 year.

Rating: Good

◆ Countertop Filters with Standard 10-Inch Cartridges

There are hundreds of different countertop units made to fit standard, interchangeable filter cartridges; most are durable, high-quality units that are relatively inexpensive. Almost all standard countertop units come with a single cartridge only, but there are a few exceptions that use two cartridges in two housings. The availability of a wide variety of standard cartridges allows consumers to customize a water purifier for specific water conditions and to purchase cartridges from a competitive array of suppliers.

AQUATIC REEF COUNTERTOP. An inexpensive single-housing unit with carbon block standard interchangeable cartridge. Can be used with any other standard cartridges, including multi-stage. Attaches to kitchen faucet, comes with its own faucet.

Price range: About $40

Replacement cartridges: $10

Cartridge life: 6 months

Rating: Good

PURE WATER PRODUCTS LLC, 77 AND 77S. A single-housing unit made with top-quality parts. Comes with the widest choice of carbon, redox, alumina, ion-exchange, and other cartridges, all for the same price. Different colors and styles of the countertop unit are also available for the same price. Comes with its own faucet. Model 77S is for candle-style cartridges, such as ceramic. Lifetime warranty on all components excluding cartridges. Top value!

Price range: About $75

Replacement cartridges: $20 to 44 depending on type

Cartridge life: Depends on type, typically 6 months to 2 years

Rating: Good

COUNTERTOP SUPERIOR WATER FILTER SYSTEM. A single-housing unit with a multistage cartridge that has a carbon filter, blocks cysts, and reduces lead. Comes with its own faucet. Top value!

NSF 53 certified

Price range: About $60

Typical Countertop Filter with Standard Cartridges

Proprietary 10-inch cartridges: About $25. Alternate cartridges may be used.

Cartridge life: 1 year claimed (6 month recommended)

Rating: Good

PURITEC CT-12. A three-stage filter within a single housing: has a microfilter for cysts, a redox stage for removing lead and chlorine, and a granular carbon stage. Unit has its own faucet.

Price range: $110–$150

Proprietary 10-inch cartridges: $60. Alternate cartridges may be used.

Cartridge life: About 3 years claimed (6 months recommended)

Rating: Good

PURE WATER PRODUCTS DOUBLE 77 AND DOUBLE 77S. A double-housing unit made with top-quality parts. Comes with the widest choice of carbon, redox, alumina, ion-exchange, and other cartridges, all for the same price. Different colors and styles of the countertop unit are also available for the same price. Comes with its own faucet. Model 77S is for candle-style cartridges, such as ceramic. Lifetime warranty on all components excluding cartridges. Flexibility and effectiveness in a low-cost unit. Top value!

Price: $134

Replacement cartridges: $20–$44 depending on type

Cartridge life: Depends on type, typically 6 months to 2 years

Rating: Good

PURITEC CTD-12. A two-housing unit. The first housing has a ceramic filter for trapping bacteria and cysts, and the second housing has a redox stage for lead and chlorine removal, plus a granular carbon stage. Unit has its own faucet.

Price range: $160–$190

Proprietary 10-inch cartridges: GAC/redox cartridge, $60; ceramic cartridge, $60. Alternate cartridges may be used.

Cartridge life: GAC/redox cartridge life, about 3 years claimed (6 months recommended); ceramic, about 3 years with normal use, should be inspected every few months and scrubbed when necessary.

Rating: Good

◆ Under-Sink Filters with Standard 10-Inch Cartridges

There are hundreds of under-sink filter systems made to fit standard, interchangeable filter cartridges; most are durable, high-quality units that are relatively inexpensive. These units come with one to four cartridge housings.

Some cartridges are single-function filters, and others come with multiple stages within a single cartridge body. The availability of a wide variety of standard cartridges allows consumers to customize a water purifier for specific water conditions and to purchase cartridges from a competitive array of suppliers.

All under-sink filters require installation and come with a separate faucet that is installed on the countertop. All prices listed here include the countertop faucet and connecting hardware.

PURE WATER PRODUCTS LLC, SINGLE UNDERSINK FILTER. A single-housing unit made with top-quality parts. Comes with your choice of one combination, carbon, redox, alumina, ion-exchange, or other cartridge, all for the same price. Uses quick-attach parts. Shipping cost is included in the price. Some flexibility in a low-cost unit. Top value!

Price: $93

Replacement cartridges: $20–$44 depending on type

Cartridge life: Depends on type, typically 6 months to 2 years

Rating: Good

Typical Under-Sink Filter with
Standard Cartridges

UNDERCOUNTER SINGLE SUPERIOR. A single-housing unit that has a carbon filter, cyst-removal filter, and reduces lead. Comes with a top-quality, multistage standard cartridge. Top value!

NSF 53 certified

Price range: $100–$120

Proprietary 10-inch cartridges: About $25. Alternate cartridges may be used.

Cartridge life: Approximately 1 year (6 months replacement recommended)

Rating: Good

PURITEC UC-12. A single-housing unit with a microfilter stage for cysts, a redox stage for reducing chlorine and lead, and a granular carbon stage for other pollutants. Comes with a high-quality cartridge.

Price range: $150–$170

Replacement cartridges (proprietary): $60. Alternate cartridges may be used.

Cartridge life: 3 years claimed (6 months recommended)

Rating: Good

CARBOSTAR UC-1. A single-housing unit with a unique, ultracompacted submicron carbon block filter that is, as of this writing, the only carbon filter that traps all pathogens, *including viruses,* in addition to all the pollutants typically removed by carbon block filters. While it is not recommended to use this filter by itself on water known to be contaminated with pathogens, it should be considered when there has been evidence of the occasional presence of cysts, bacteria, or viruses, or when the presence or absence of pathogens is unknown. Cartridge has a built-in shutoff valve that kicks in when the filter is exhausted.

Certified by a laboratory approved by the USDA, FDA, and CDC

Price range: $204, including one spare cartridge

Replacement cartridges (proprietary): $85 each. Cartridges fit all 10-inch standard housings, and alternate cartridges may be used in the Carbostar housing.

Cartridge life: Manufacturer claims 2 years under normal conditions. Although other carbon cartridges should be replaced every 6 months, an extended life for this cartridge may be justified due to its special nature.

Rating: Good

PURE WATER PRODUCTS LLC, DOUBLE UNDERSINK FILTER. A double-housing unit made with top-quality parts. Comes with two cartridges for removal of chlorine, lead, cysts, and so on, or select from a variety of carbon, redox, alumina, ion-exchange, and other cartridges, all for the same price. Employs quick-attach connections. Shipping cost is included in the price. Flexibility and effectiveness in a low-cost unit. Top value!

Price: $151

Replacement cartridges: $20–$44 depending on type

Cartridge life: Depends on type, typically 6 months to 2 years

Rating: Good

UNDERCOUNTER TWIN SUPERIOR. A two-housing unit. The first stage is a combination granular carbon and redox filter for removal of lead and other pollutants; the second stage is a carbon block filter for removal of cysts and further removal of other pollutants. Good value!

NSF 53 certified

Price range: $140–$195

Replacement cartridges (proprietary): About $25. Alternate cartridges may be used.

Cartridges life: Approximately 1 year (6 months recommended)

Rating: Good

PURITEC UCD-12. A two-housing unit. The first housing has a pleated membrane microfilter for trapping bacteria and cysts. A three-stage filter within the second housing has an additional microfilter, a redox stage for lead and chlorine removal, and a granular carbon stage.

Price range: $200–$220

Replacement cartridges (proprietary): Sediment microfilter cartridge, $16; GAC/redox cartridge, $60. Alternate cartridges may be used.

Cartridge life: Sediment microfilter, 6 to 12 months claimed (6 months recommended); GAC/redox cartridge, 3 years claimed (6 months recommended)

Rating: Good

PURITEC UCD-12C. A two-housing unit. The first stage is a ceramic filter for trapping bacteria and cysts; the second stage is a combination filter, redox for removal of chlorine and lead and granular carbon for other pollutants.

Price range: $250–$270

Replacement cartridges (proprietary): Ceramic cartridge, $60; GAC/redox cartridge, $60. Alternate cartridges may be used.

Cartridge life: Ceramic cartridge, about 3 years with normal use (should be inspected every few months and scrubbed when necessary); GAC/redox cartridge, about 3 years claimed (6 months recommended)

Rating: Good

PURE WATER PRODUCTS LLC, TRIPLE UNDERSINK FILTER. A triple-housing unit made with top-quality parts. Comes with three standard cartridges for removal of chlorine, lead, cysts, bacteria, and other pollutants, or select from a variety of carbon, redox, alumina, ion-exchange, and other cartridges, all for the same price. Employs quick-attach connections. Shipping cost is included in the price. Provides more flexibility and effectiveness in a low-cost unit. Top value!

Price: $199

Replacement cartridges: $20–$44 depending on type

Cartridge life: Depends on type, typically 6 months to 2 years

Rating: Very Good

PURITEC UCT. A three-housing unit. The first housing has a pleated membrane microfilter for trapping bacteria and cysts. A three-stage filter within the second housing has an additional microfilter, a redox stage for lead and chlorine removal, and a granular carbon stage. The third housing can be ordered with an alumina cartridge for fluoride removal, a resin cartridge for nitrates removal, or a carbon block cartridge for additional general reduction of pollutants.

Price range: $300–$320

Replacement cartridges (proprietary): Sediment microfilter, $16; GAC/redox, $60; fluoride or nitrates, $70 each. Alternate cartridges may be used.

Cartridge life: Sediment microfilter, 6 to 12 months claimed (6 months recommended); GAC/redox cartridge, about 3 years claimed (6 months recommended); fluoride and nitrates cartridges, 6 months (nitrates cartridge can be washed and used several times)

Rating: Good

PURE WATER PRODUCTS LLC, QUADRUPLE UNDERSINK SUPER FILTER. A four-housing unit made with top-quality parts. This is the ultimate combination filter system for both flexibility and effectiveness. Comes with four standard cartridges for removal of chlorine, lead, cysts, bacteria, and other pollutants, or select from a variety of carbon, redox, alumina, ion-exchange, and other cartridges, all for the same price. The difference between this unit and the triple-housing unit by the same manufacturer is that this one removes a higher percentage of pollutants. Employs quick-attach connections. Shipping cost is included in the price. A great value for the performance!

Price: $229

Replacement cartridges: $20–$44 depending on type

Cartridge life: Depends on type, typically 6 months to 2 years

Rating: Very Good

Reverse Osmosis Purifiers

◆ Miniature, Faucet-Mount Reverse Osmosis Purifiers

There are only a few manufacturers of miniature, faucet-mount, reverse osmosis units. They are relatively inexpensive, simple to use, and fairly effective. Because the RO membranes are so small, the water is produced in a slow trickle, and since there are no tanks to store the treated water, it must be gathered in a container in the sink, under the RO unit. Compared to faucet-mount filters, they are more expensive but they do a slightly better job of removing pollutants than the best-quality faucet filters. Like snap-on faucet filters, faucet ROs can be stored in a refrigerator when not in use.

PUROSMART FAUCET REVERSE OSMOSIS. Produces 15 gallons per day. Includes a separate carbon block filter as well as an RO membrane.

Price range: $100–$130

Replacement components (proprietary): RO membrane, $50; carbon block cartridge, $7

Component life: RO, 1 year; cartridges, 3 months

Rating: Good

NIMBUS WATERMAKER MINI. Produces 8 gallons per day. Employs a sediment pad and granular carbon filter.

NSF 58 certified

Price range: $90–$100

Replacement components (proprietary): RO membrane, $50

Component life: 9 months to 1 year with average use

Rating: Good

◆ Countertop Reverse Osmosis Purifiers

Countertop RO units are full-size, multistage systems with performance that approaches or equals that of under-sink RO units, but

131

Typical Miniature Faucet-
Mount Reverse Osmosis Unit

they snap onto a kitchen faucet and can be moved when not in use. Some units have their own faucet built in and some attach to the kitchen sink faucet.

Note: All RO systems that employ a UV stage are in the UV purifier listings, not in the RO listings.

TGI (TOPWAY GLOBAL, INC.) CT-445. An attractive countertop unit with its own faucet connects to the kitchen faucet with a snap-on fitting. It has four stages: sediment filter, granular carbon filter, an RO membrane, and a final granular carbon stage. It produces 1.5 gallons per hour, and the storage tank holds 1.75 gallons. Input water flow is automatically shut off when the tank is full.

Price range: $250–$290; optional pump for low water pressure, $120

Replacement components (proprietary): Sediment, $9–$13; GAC, $8–$14 (2 required); RO membrane, $57

Component life: Sediment, 6 months; GAC, 6 months; RO membrane, about 2 years

Rating: Very Good

TGI (TOPWAY GLOBAL, INC.) CT-475R. Same as the TGI CT-445 (above) but with a larger RO membrane, which yields 3 gallons per hour and a larger storage tank (2.5 gallons).

Price range: About $400

Replacement components (proprietary): Sediment, $9–$13; GAC, $8–$14 (2 required); RO membrane, $57

Component life: Sediment, 6 months; GAC, 6 months; RO, about 2 years

Rating: Very Good

PURE WATER PRODUCTS CTRO-A. This is a bare-bones unit with exposed cartridge housings and tubing. But if you need a full-size RO unit that's portable (for example, to take to a vacation home), that does a good job of removing most pollutants, and that you can snap onto the sink faucet as needed, this one should be considered. This three-stage unit usually comes with two carbon block filters and an RO membrane that makes about 2 gallons per hour. It doesn't have a tank, so you must use your own container, and a sink drain is needed. An advantage of this unit is that you can order it with any of a variety of standard 10-inch cartridges and/or RO membranes to suit your water conditions, and you can purchase replacements from a range of suppliers.

Price range: $170–$200

Typical Countertop Reverse Osmosis Unit

Replacement components (list of cartridges is partial): Sediment, $7–$20; GAC, $10–$40; carbon block, $15–$30; ceramic, $33–$46; deionization, $27; nitrate, $35; arsenic-specific, $60. This unit may be ordered with a variety of RO membranes that produce from 0.4 gallons per hour to 4 gallons per hour and cost $45–$80.

Component life: Cartridges vary depending on type; RO membrane, 2 to 4 years

Rating: Good

CRYSTAL QUEST COUNTERTOP REVERSE OSMOSIS. This is a bare-bones unit with exposed cartridge housings and tubing. But if you need a full-size RO unit that is portable (for example, to take to a vacation home), that does a good job of removing most pollutants, and that you can snap onto the sink faucet as needed, this one should be considered. This is a three-stage unit: sediment filter, granular carbon filter, and an RO membrane that makes about 2 gallons per hour. It doesn't have a tank, so you must use your own container, and a sink drain is needed.

Price range: $170–$200

Replacement components (proprietary): Sediment/cyst, $15; GAC, $15; RO membrane, $45–$55

Component life: Cartridges, 6 months; RO membrane, 2 to 4 years

Rating: Good

◆ Under-Sink Reverse Osmosis Purifiers

There are hundreds of brands of under-sink RO systems. These are typically the best-performing RO units in that they offer more stages of treatment than smaller countertop models. Also, many under-sink ROs employ booster pumps that maintain a higher water pressure, which allows the RO membrane to perform at its peak. When booster pumps are used, some run on electricity and some are powered by water pressure. All under-sink ROs require instal-

Pure Water Products CTRO-A
Reverse Osmosis System

lation, and almost all come with a separate faucet that is installed on the countertop. All models recommended here come with their separate faucet and all connecting hardware. Some rapid flow ROs do not require storage tanks.

Note: All RO systems that employ a UV stage are in the UV purifier listings, not in the RO listings.

RIOFLOW USRO5-50. A low-cost, high-performance system with five stages: a sediment filter, two separate carbon block stages, an RO membrane, and a final granular carbon stage. It makes 2 gallons per hour (3 gallons per hour and 4 gallons per hour systems are available at slightly higher costs). The tank holds 3.2 gallons and has an automatic shutoff. A top value in under-sink RO systems!

Membrane and filters NSF certified; overall system not certified

Price range: About $180

Replacement components (standard cartridges): Sediment, $3; carbon block, $16 (2 required); in-line GAC cartridge, $12; RO membrane, $44

Component life: Prefilters, 6 months; postfilter, 1 year; RO, 2 years

Rating: Very Good

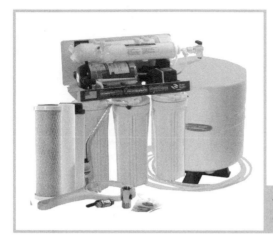

Typical Under-Sink Reverse Osmosis Unit

TGI (TOPWAY, GLOBAL, INC.) WIN445. A low-cost, high-performance system with four stages: sediment filter, block carbon filter, RO membrane, and a final granular carbon stage. It makes 2 gallons per hour. The tank holds 4 gallons and has an automatic shutoff. A top value in under-sink RO systems.

WQA certified

Price range: About $200

Replacement components (standard cartridges): Sediment, $12–$16; carbon block, $15; RO membrane, $60; in-line GAC postfilter, $15

Component life: Prefilters, 6 months; postfilter, 1 year; RO membrane, 2 years

Rating: Very Good

PUREGEN ERO-535. A low-cost, high-performance system with five stages: sediment filter, granular carbon filter, carbon block filter, RO membrane, and a final granular carbon stage. Makes 1.5 gallons per hour (2 gallons per hour and 3 gallons per hour systems are available at slightly higher costs). The tank holds 3.2 gallons and has an automatic shutoff. A good value in an under-sink RO system.

WQA certified

Price range: $170–$250

Replacement components (standard cartridges): Sediment, $6; carbon, $9; carbon block, $10; RO membrane, $43; in-line carbon postfilter, $10

Component life: Filters, 6 months; RO, 2 years

Rating: Very Good

GE MERLIN RO SYSTEM. A full-size, multistage unit that delivers a very rapid flow of treated water and is especially advantageous where larger quantities are needed. It produces an average of 0.75 gallons per minute, or 45 gallons per hour. There is no pump, and since there's no tank, you must provide your own collection bottles. The unit employs a granular carbon prefilter, an RO membrane, and a granular carbon postfilter.

NSF 58 certified and WQA Gold Seal

Price range: $400–$450

Replacement components (proprietary): Carbon filters, $20–$25 each; RO membrane, $95

Component life: Carbon filters, 6 months; RO membrane, 2 to 4 years

Rating: Very Good

GE Merlin Reverse Osmosis

Distillers

◆ Countertop Manual Distillers

Countertop distillers are portable, fairly compact units that have their own container, often a fitted pitcher. You have to manually pour tap water into them and start them, but they shut off automatically when full. After you fill them, they can be moved out of the way; that's a good thing, as they take considerable time to distill the water, all the while generating heat, as well as noise from the cooling fan.

If you use hard water in a distiller, scale will build up in the boiling chamber; with soft water there is little or no buildup. Because scale buildup may occur, it's important to have easy access to the boiling chamber for cleaning. The scale that builds up is hard, and removing it is made much easier if you use a special descaling solution sold by distiller suppliers and plumbing stores.

While a few countertop distillers come with glass collection bottles, most come with polycarbonate plastic bottles, the same type of plastic commonly used by water delivery services for their water. A few distillers come with plastic collection bottles made of PET plastic, which is used for essentially all plastic containers for bottled drinks. PET plastic has been studied extensively and is considered safe. Polycarbonate bottles, however, have been found to leach traces of chemicals that can be harmful to health, albeit in quantities judged too small to have any effect. If the distiller you choose has a polycarbonate bottle, to be on the safe side you may consider transferring the distilled water to a glass bottle as soon as convenient or refrigerate the bottle to reduce possible leaching.

Water distillation is a slow process, and while the small capacity of a countertop unit is probably adequate for most small families, if you have a large household you'd have to run it often.

Finally, because distillers contain electrical components and because they operate at high temperatures, it is essential that they be dependable and that parts and service be readily available. The following units fit these criteria.

MEGAHOME/ECOPURE/LOVE COUNTERTOP DISTILLER. Sold under various brand names, this is the most widely sold distiller in the world. It has many quality features in spite of its low price. It is an air-cooled, lightweight portable unit that comes with a 1-gallon, polycarbonate collecting bottle. It produces 1 gallon of water in 5 hours. The water is then passed through a small granular carbon filter. The filter connection to the distiller and the collection bottle is sealed, so there is minimal risk of contamination by air. This distiller has a stainless steel boiling chamber that is easily accessible for cleaning, and the electrical heating element is under the boiler, so it does not have to be cleaned. Automatic shutoff.

Electricity usage: 580 watts

Price range: $100–$175

Replacement filters: $1–$3

Filter life: 1 to 2 months

Rating: Excellent

WATERWISE 4000. This is a more powerful version of the Megahome countertop distiller (above) with the same features but produces 1 gallon every 4 hours.

WQA, CSA (Canada), and CE (European Union) certified

Electricity usage: 800 watts

Waterwise 4000 and Megahome/Ecopure/Love Distillers

Kenmore 34480 and Waterwise
8800 Distillers

Price range: $240–$330

Replacement filters: about $5

Filter life: 1 to 2 months

Rating: Excellent

KENMORE 34480. An inexpensive, high-quality, air-cooled distiller with side-by-side removable, stainless steel boiling chamber (for easy cleaning) and 3-quart polycarbonate collecting bottle. The heating element is under the boiler so it does not have to be cleaned. Automatic shutoff. It produces 3 quarts every 4.5 hours (or 1 gallon every 6 hours). The distilled water is passed through a small carbon filter in the lid of the collection bottle.

NSF 62 certified

Electricity usage: 500 watts

Price range: $100–$140

Replacement filters: $3–$6

Filter life: 1 to 2 months

Rating: Excellent

WATERWISE 8800. A more powerful version of the Kenmore 34480 (above). It has the same features but produces 1 gallon every 4 hours. This distiller also has an electronic programmable timer to start distillation at any hour—a very useful feature. It includes a 1-gallon polycarbonate bottle.

NSF 62 certified

Electricity usage: 800 watts

Price range: $300–$400

Replacement filters: $3–$5

Filter life: 1 to 2 months

Rating: Excellent

PURE WATER MINI-CLASSIC. This entirely stainless steel counter-top unit is very rugged and, at 24 pounds, is heavier than the lightweight distillers described above. It has a removable, stainless steel boiling chamber and stainless steel condensing coils. It produces 1 gallon every 5 hours and fills a 1-gallon glass bottle. The filter housing is also stainless steel and fits above the collection bottle; there is no water contact with plastic.

Electricity usage: 800 watts

Price range: $600–$700

Replacement filters: $2–$5

Filter life: 1 to 2 months

Rating: Excellent

Pure Water
Mini-Classic Distiller

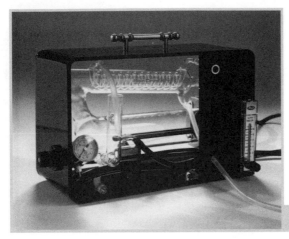

Rain Crystal Distiller

RAIN CRYSTAL BY SCIENTIFIC GLASS CO. Even stainless steel leaches minute amounts of impurities into water. The Rain Crystal is a unique water-cooled, lightweight, portable distiller whose innards are all made of high-quality, handblown laboratory glass. Lab glass is used because it is very strong, so in spite of being made of glass it's quite tough. Lab glass is also the most inert of all materials used for distillation, so there's no chance of any impurities leaching from the distiller components. Not only does the Rain Crystal produce very pure water, it's also fun to watch the transparent distillation process.

This little handcrafted distiller has been around for a long time and is one of the few water-cooled distillers made for home use. It is almost completely silent when running, and it generates very little heat. However, like all water-cooled distillers, it uses more water than batch-type distillers—5 gallons for every 1 gallon it makes.

The Rain Crystal snaps on to a kitchen faucet and drains into the sink. You don't fill it, you simply adjust water flow for proper operation and turn it on. You have to use your own collecting container, and you set a timer to shut it off. It makes 1 gallon every 3 hours.

The reason its rating is "Good to Excellent" is because it doesn't come with carbon filtration, so there's a possibility that some volatile organic chemicals (VOCs) may still be in the water after distillation. For excellent performance, simply place a small bag of granular carbon (GAC) between the output and the collecting container (for example, in a funnel). Refrigerate the granular carbon when not in use. These granular carbon bags, or pouches, are inexpensive and are available from most distiller suppliers. Also, be aware that there is no access to the innards of this distiller. It should be used with softened water to prevent scale buildup. If softened water isn't available, it must be periodically cleaned with a descaling solution.

Price range: About $700. A good value for a laboratory glass distiller. And it is a terrific conversation piece!

Rating: Good to Excellent

◆ Automatic Distillers

Automatic distillers are permanently connected to a cold water line. A device in the storage tank determines when the level is low and turns on the distiller to produce more water. Thus, purified water is always available. Some models can be batch-type; that is, they fill up with water and produce one batch of treated water at a time. These are always air-cooled with a fan. As with manual distillers, batch-type units generate some fan noise and heat. Other automatic distillers are water-cooled and have a continuous flow of water into the distiller when operating; these are less common and only a few models are available.

Automatic distillers either have their own spigot on the storage tank or pump water to one or more faucets located in the kitchen or other rooms. When these units are connected to external faucets, a demand pump is included; the pump turns on only for the few seconds that water is being drawn from a faucet.

Automatic distillers typically can be ordered with a range of storage tank sizes, from modest to quite large, depending on the amount of water used.

Scale needs to be cleaned from all distillers periodically, especially when hard water is used. It's important to have easy access to the boiling chamber, where scale buildup is most prevalent. Unlike portable countertop distillers, which have removable collecting pitchers, the storage tanks of automatic distillers are permanently in place, so it's important to also have good access to them for occasional cleaning. Finally, because automatic distillers contain electrical components and because they operate at high temperatures, it is essential that they be dependable and that parts and service be readily available. The following units fit these criteria.

WATERWISE 7000. This high-quality stainless steel distiller, comes as a countertop model or a floor-standing model depending on the size of the holding tank. It has proved to be very dependable. It also has a number of desirable features. It has a removable stainless steel boiling chamber for easy cleaning. This distiller is one of the best in terms of access to the boiling chamber and holding tank. The condensing coils are also stainless steel. It can be ordered with either a front or side spigot for convenient positioning, and it has a sight gauge for visually determining the level of water in the holding tank.

The unit has been carefully designed for safety; no visible controls are accessible to children, and there are no exposed hot surfaces. It produces 1 gallon every 3 hours and comes with a small, granular carbon postfilter.

The countertop unit comes with a 3-gallon holding tank; the floor-standing units come with 8- or 12-gallon holding tanks. A demand pump is optional for one or more countertop faucets and/or connection to an ice maker. Another option is a standard 10-inch carbon block prefilter. With the prefilter, this unit achieves the highest level of water purification of any type of system. Note that the "Excellent++" rating applies when used with the optional carbon block prefilter.

WQA, UL, CSA (Canadian), and CE (European Union) certified

Electricity usage: 1,200 watts

Price range: 3-gallon model, $1,300–$1,600; 8-gallon model, $1,500–$1,750; 12-gallon model, $1,750–$2,000; optional

Waterwise 7000 Series and Dol-Fyn Series Distillers

demand pump and connections, $350–$400; prefilter, $45–$55

Replacement components: Postfilter, $6–$7; prefilter cartridge $25, interchangeable with all standard 10-inch cartridges

Filter life: Prefilter, about 1 year; postfilter, about every 3 months

Rating: Excellent to Excellent++

AQUA TECHNOLOGY/DOL-FYN. This series is made by the same manufacturer as the Waterwise 7000 series (above) and is basically the same. However, the Aqua Technology series all come with two standard 10-inch prefilters: sediment and carbon block. No postfilter is used. This series represents the highest level of water purification of any type system. The Dol-Fyn No. 1 comes with a 4-gallon tank; the Dol-Fyn No. 2 has an 8-gallon tank; and the Dol-Fyn No. 3 has a 12-gallon tank.

Price range: 3-gallon model, about $1,000; 8-gallon model, about $1,200; 12-gallon model, about $1,400; optional demand pump and connections, $160

Replacement components: Prefilters, $25 (pair)

Component life: Prefilters, 1 year

Rating: Excellent++

GLACIER D-3. This is a water-cooled, automatic distiller, which gives it some advantages over air-cooled distillers. Because the unit generates little or no heat or humidity to the surroundings, it can be installed in an enclosed space such as under the sink. One advantage is higher efficiency; the heat created to boil water is used to preheat incoming water, which results in a lower electricity cost per gallon produced. Another advantage is that when the distilled water enters the storage tank it has been cooled, unlike air-cooled distillers where the distilled water is hot and takes some time to cool. And the unit operates silently, with no fan noise. However, all water-cooled distillers require more input water to produce distilled water than do air-cooled distillers; typically, 5 gallons of input water for every gallon of distilled water produced.

The D-3 is stainless steel and has a 5-gallon storage tank. It produces approximately 1 gallon every 2.5 hours. The unit features an automatic boiling chamber flush and a self-sterilizing storage tank. It comes with countertop faucet, hardware, and a demand pump as standard. Vents are included to help remove VOCs. There is a screen-type sediment prefilter that does not require replacement and a carbon postfilter. A mineralizing postfilter is available as an option for those who prefer to have the distilled water re-mineralized.

Electricity usage: 1,000 watts

Price range: About $2,200

Replacement components: Carbon postfilter, optional mineralizing filter

Filter life: 1 year

Rating: Excellent

Waterwise 1600 Nonelectric Distiller

◆ Nonelectric Distillers

There are a few distillers built to operate from a heat source other than electricity. The heat source can be a gas stove, a propane grill, a campfire, and so forth (or an electric stove). They are used when electricity is not available or as a source of purified water in an emergency.

WATERWISE 1600. This entirely stainless steel unit with no moving parts is about the size of a large cooking pot. With an adequate heat source, it produces 1 gallon every 1.5 hours. There is no auxiliary filter. It comes with a small, separate digital timer that can remind you to turn off the heat, but it isn't connected to the distiller.

Price range: $270–$370; optional 3-quart stainless steel closed collection tray, $50–$60; 3-gallon polycarbonate collection bottle with 36 inches of tubing, $25–$30

Rating: Very Good

◆ Solar Distillers

A solar distiller seems like a good idea. There is no energy cost because the sun provides it, and it is a natural process. Solar distillers are typically made in the form of a large, shallow trough that holds water, with a glass top facing the sun. Sunlight penetrates the glass and heats the water, which evaporates and condenses on the glass top. The condensate is then routed through a tube to an external container.

One of the main attributes of electrical distillers is that they kill all pathogens. However, solar distillers can't do this; they don't boil the water, and they aren't sealed. They might more properly be called "solar evaporators."Furthermore, their large, warm, moist surface areas—located out of doors—can collect a variety of insects and microorganisms that are transferred into the collected water.

Solar distillers are also quite slow and dependent on weather conditions. They can be of use as an emergency drinking water supply or possibly a regular drinking water supply, if the collected water is disinfected by a separate process.

Ultraviolet Purifiers

♦ Countertop Ultraviolet Purifiers

Countertop ultraviolet systems typically come with one or more filters. They either snap onto the kitchen faucet or come with their own separate faucet.

PURITEC CT-UV COUNTERTOP. A single-housing unit that contains both the UV lamp, which kills bacteria and viruses, and a carbon block filter for cysts and other pollutants. The unit has its own faucet. If used with a well, spring, or other untreated water source, a separate sediment prefilter must also be used. Top value!

Price range: About $190

Replacement components: Proprietary carbon block cartridge, $25; UV lamp, $40 *Component life:* Cartridge, 1 year claimed (6 months recommended); UV lamp, 1 year

Rating: Good

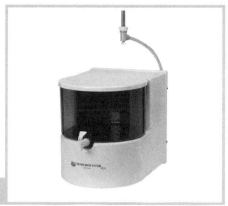

Typical Countertop Ultraviolet Unit

AQUA SUN CT100 COUNTERTOP. A single-housing unit that contains both the UV lamp, which kills bacteria and viruses, and a carbon block filter for cysts and other pollutants. The unit has its own faucet. If used with a well, spring, or other untreated water source, a separate sediment prefilter must also be used. Top value!

Price range: About $200

Replacement components (proprietary): Carbon block cartridge, $40; UV lamp, $60; quartz sleeve replacement (if damaged), $55

Component life: Cartridges, 1 year claimed (6 months recommended); UV lamp, 1 year

Rating: Good

TGI (TOPWAY GLOBAL, INC.) CT475P-UV. This unit uses a highly effective combination of reverse osmosis, filters, and UV radiation to remove essentially all potential pollutants. It is an attractive countertop unit with its own faucet, and it connects to the kitchen faucet with a snap-on fitting. It has five stages: sediment filter, granular carbon filter, RO membrane, an additional granular carbon stage, and a UV stage. It produces 2.5 gallons per hour, and the storage tank holds 1.25 gallons. When the tank is full, incoming water is automatically shut off.

Price range: About $540

Replacement components (proprietary): Sediment, $9–$13; GAC, $8–$14 (2 required); RO membrane, $57; UV lamp, $78

Component life: Sediment, 6 months; GAC, 6 months; RO, about 2 years; UV lamp, 1 year

Rating: Excellent

Under-Sink Ultraviolet Purifiers

VERTEX UV-2. A compact filter and UV combination with two standard housings and three stages: sediment/cyst filter, carbon block filter, and UV. Comes with a separate faucet and all connecting hardware.

Price range: About $275

Replacement components: Sediment/cyst, $12; carbon block, $15; UV lamp, $65

Component life: Filters, 6 months; UV lamp, 1 year

Rating: Good

PURA SERIES OF UNDER-SINK UV PURIFIERS. The Pura series of UV purifiers come in combination with one or more carbon filters in compact, standard housings. They cost more than many other UV systems, but they are simple and very dependable.

Price range: $290–$520

Replacement components: Carbon cartridges, $25; sediment cartridges, $5–$10; UV lamp, $50

Component life: Cartridges, 6 months; UV lamp, 1 year

Rating: Good

PUREGEN EROUV-600. This is a highly effective system that will remove essentially all potential pollutants. It has six stages: sediment, granular carbon, carbon block, RO membrane, inline granular carbon postfilter, and ultraviolet. It makes 1.5 gallons per hour (2 gallons per hour and 3 gallons per hour systems are available at slightly higher costs). The tank holds 3.2 gallons and has automatic shutoff. It comes with a separate faucet and all connecting hardware. Top value!

WQA certified

Price range: $280–$300

Replacement components (standard cartridges): Sediment, $6; carbon, $9; carbon block, $10; in-line GAC postfilter, $10; RO membrane, $43; UV lamp, $100

Component life: Standard cartridges, 6 months; in-line GAC filter, 1 year; RO membrane, 2 years; UV lamp, 1 year

Rating: Excellent

TGI (TOPWAY GLOBAL, INC.) 625U. This is a highly effective system that will remove essentially all potential pollutants. It has six stages: sediment, granular carbon, carbon block, RO membrane, in-line granular carbon postfilter, and ultraviolet. It makes 2 gallons per hour. The tank holds 4 gallons and has automatic shutoff. It comes with a separate faucet and all connecting hardware. Top value!

WQA certified

Price range: About $350

Replacement components (standard cartridges): Sediment, $12–$16; carbon, $15; carbon block, $15; GAC filters, $15; RO membrane, $60; UV lamp, $60

Filter life: Standard cartridges, 6 months; in-line GAC filter, 1 year; RO membrane, 2 years; UV lamp, 1 year

Rating: Excellent

Other Types of Purifiers

● Ozonators

The basics of ozonation are described at the end of chapter 7. While most ozonation systems for home use are for whole-house removal of iron, manganese, and other staining minerals, they are also used to treat drinking water.

Of the five categories of water pollutants—pathogens, toxic minerals and metals, toxic organic compounds, toxic inorganic compounds, and radioactive substances, ozone oxidizes most of them and renders them harmless. After the ozonation process, filters remove the remaining residue. Ozonation in combination with filtration is especially effective at killing and removing all harmful pathogens.

Ozonators require electricity but only a minimal amount. Ozonators themselves have no components requiring periodic replacement and, as with distillation, there is no reduction in their performance over time. Because ozone is very chemically aggressive and can deteriorate components of a water purifier, it is very important to select an ozonator from an established, reputable company.

BELLE AQUA 300. This ozonator can be mounted on a countertop or under the sink or be used as a portable unit. It has its own faucet. It is a combination ozone generator and a four-stage postfilter. The filter stages are sediment, granular carbon, resin for metals, and carbon block. Because the filter is downstream of the ozone, it receives disinfected water and thus isn't subject to internal growth of microorganisms. Comes with a faucet and connection hardware.

The filter is NSF 53 certified, but certifications for ozonators for treating drinking water haven't been established.

Price range: About $400

Replacement components: $50

Component life: 1 year

Rating: Very Good

◆ Air-to-Water Purifiers

Purifiers that extract water from the air and then further purify the water produced are now under development, and some are already on the market. They work best in warm, humid climates.

EVEREST ATMOSPHERIC WATER GENERATOR E10. This is a floor-standing hot and cold water unit that looks like a traditional water dispenser. It extracts water from the air, filters it, and passes it through a series of extensive treatment stages before storing the water in holding tanks. If ambient air conditions do not supply enough water condensate (because of lack of humidity or cool temperatures), a backup water line supplies water for the same stages of treatment.

The specific process is as follows: Intake air is passed through an electrostatic filter and an air microfilter and is then routed through condensing coils. The water then passes through a UV lamp, a sediment filter, a patented sterilizing filter, a carbon block filter, a membrane ultrafilter, and, finally, a second UV lamp. The finished water is then routed to hot and cold holding tanks.

The system can produce up to 13 gallons of treated water per day under ideal conditions (temperature 90°F with 85 percent humidity) but only about 2 gallons per day when conditions are more moderate (70°F with 45 percent humidity). However, if treated water in the cold water tank falls below 2 gallons, the backup water line opens and water is treated from the connected source.

Because of the air filtration and extensive, redundant water treatment stages, the system produces very pure water totally free of pathogens. The system is WQA and CE (European Union) certified and the manufacturer claims it exceeds all NSF and UL standards. Contact the Air to Water Company for certification and cost details.

◆ Latest Developments in Pathogen Filtration

Distillation, ultraviolet radiation and ozonation *kill* harmful pathogens. Reverse osmosis membranes impede them and flush them away to a drain. Filtration captures them within the filter medium and prevents them from passing downstream. Because the pathogens are not killed, RO and filters have to achieve very high manufacturing standards to make sure no pathogens get past the treatment unit. In the past the only ROs or filters able to do this were those for research and medical applications, etc. Recently some companies have announced water purifiers for home use that claim to meet this high standard. For these units, certification by recognized testing organizations is essential.

While UV and ozonation kill most pathogens, some can slip through. And while distillation kills all pathogens, it requires lots of electricity. So the advantages of a certified RO/filter microbial purifier are that it traps all pathogens, it doesn't require electricity, and it is silent and doesn't generate heat.

One disadvantage of this type of purifier is that because of the high manufacturing standards required, its cost (so far) is high. Other potential disadvantages are that the system must be carefully monitored for performance, its replaceable components must be replaced on time, and it must be periodically sanitized.

Nevertheless, there are applications where a mechanical microbial purifier has some clear advantages over other purification methods. As of this writing, only one water treatment system of this kind, a combination of RO and filters, has been certified as a microbial water purifier for home use, and it is described below.

PALL CORP. PUREFECTA. This is a 5-stage, undersink purifier consisting of a sediment filter, a self-cleaning RO stage, a virus filter, a carbon block filter and a bacteria filter. It includes a countertop faucet, a filter-life indicator, and a fail-safe alert if performance should decrease. It produces 0.8 gallons per hour and has a special, circulating storage tank to keep treated water fresh. Can be connected to an icemaker and/or other outlets. There are some restrictions on use with certain water conditions, such as maximum levels of hardness, iron and TDS (contact manufacturer for details). The manufacturer claims to have the longest warranty in the industry. The company is a large, established supplier of laboratory, medical, and other technical equipment.

This is the first mechanical system to be EPA approved as a microbial purifier. The system is NSF P231, NSF 42, NSF 53, and NSF 58 certified. WQA and UL certified as a microbial purifier.

Price range: About $1,600

Replacement components: Contact manufacturer for details on replacements and periodic sanitization requirement

Rating: As of this writing, insufficient time in field to evaluate

◆ Whole-House Systems

Water treatment systems that treat only drinking and cooking water are commonly called POU (point-of-use) systems. Water systems that treat water in the entire house are called POE (point-of-entry) systems. Water softeners are an example of POE systems, as are large tanks filled with various filtration media. In some instances, especially where radon gas, volatile organic chemicals (VOCs), or pathogens are in the water supply, it is desirable to have safe drinking (and breathing and washing) water at all water outlets in a house.

POE systems for health-related treatment can consist of one or more large canisters or tanks filled with granular activated carbon or other filter media. They can be simple flow-through tanks or timer-controlled tanks with automatic backwashing. Simple flow-through canisters and tanks for whole-house water treatment can cost from around $300 to $800. Automatic backwashing systems typically cost from $1,100 up to several thousand dollars. For removing chlorine, VOCs, or radon gas, a single, flow-through tank with granular carbon will suffice. If bacteria or viruses are a concern, a POE UV system or an ozonation system is required. If cysts are a concern, a large carbon block filter or an ultrafiltration filter is indicated.

Type in "whole house water filters" in any Internet search engine to find suppliers.

ZENON HOMESPRING CENTRAL WATER FILTRATION SYSTEM. This whole-house filter system provides highly purified water to all water outlets. It employs an upright floor tank that contains carbon media and an ultrafiltration membrane that traps pathogens, including all bacteria, all cysts, and some viruses. The system uses an electric controller for automatic flushing, but filtration will continue in the event of a power loss. Manual flushing is also available during a power outage. The ultrafiltration membrane should typically last 10 or more years on treated municipal water, and 5 to 10 years on untreated water, at which time the entire system needs to be replaced. An annual service inspection is recommended.

The system has NSF 42 and 53 certifications for removal of pollutants. It is not certified as a microbiological purifier although it does remove almost all potential pathogens.

Price range: $3,000–$4,000

Carbon replacement: $150–$200, typically once a year

◆ Filters for Very Dirty Water

Those of you who are on private water systems and have dirty water have probably been replacing sediment filters for years. A good alternative is a filter called the Spin-Down. Dirt in the water is separated by centrifugal action and collects at the bottom of the filter. You can see the dirt accumulation through the clear housing. When the dirt builds up, you simply open a valve at the bottom and it flows out. There is no downtime—you don't have to turn off the water to clean the unit. A reusable filter screen is occasionally removed for washing. You may still have to use conventional sediment filters, but you won't have to replace them as often. For online information on this product, type in "Rusco water filters" on any Internet search engine.

CHAPTER 12

Specialty Products
and Accessories

Reverse Osmosis Monitoring Meters

RO membranes gradually deteriorate with use, but you can test their performance using a TDS meter. (TDS stands for total dissolved solids). A TDS meter actually measures the amount of dissolved minerals in water (by measuring electrical conductivity), but the percentage of minerals removed is an indirect way of measuring the removal of *all* impurities, thus TDS provides an indirect but fairly accurate measure of the overall performance of the membrane. In general, if an RO membrane shows less than an 80 percent removal rate, it should be replaced. (Note to electricity buffs: Don't try using a common resistance meter for this task—it won't give you an accurate reading.) If your RO water purifier does not have a meter to check the performance of the RO membrane, you should get a portable meter. (Leased RO systems should be serviced and checked by the company at least every six months.)

Bottles, Dispensers, and Pumps

Most bulk water bottles now sold are made of polycarbonate, a clear, blue-tinted plastic. They're available in 2-, 3-, and 5-gallon sizes. These are strong, long-lasting, and relatively safe. However, they have been found to leach trace amounts of a chemical that can affect hormones in the body. If polycarbonate bottles are used for dispensing/storing water, it is recommended they not be used with demineralized water which is chemically aggressive and can cause more leaching. (Demineralized water is water that has been treated by distillation, deionization, or reverse osmosis.) Glass bottles are generally a better choice, especially for demineralized water. They're available in 1- and 5-gallon sizes, and they can be ordered and shipped to your residence. As of this writing, typical costs for bottles are $15 to $20 per 5-gallon bottle, plus $10 to $20 for shipping. Costs for smaller bottles range from $5 to $10.

There are several models of miniature pumps available that fit onto the necks of water bottles. These enable you to draw water from a bottle that is standing upright, so you don't have to tip the bottle for pouring. Manually operated pumps have a plunger that you move up and down, like a bicycle air pump. Battery-driven pumps do the work for you. Ceramic and plastic dispensers that support upside-down water bottles are also available.

For online ordering of bottles, dispensers, and pumps, type any of these terms into any Internet search engine: "Custom Pure Water Store," "Waterzilla," or "Waterstill." These items are also sold at stores that sell bulk water and/or water purifiers.

Shower Filters

Although this book is about drinking water, showers and baths also expose the body to tap water, along with its pollutants. There are three potentially harmful, volatile (evaporate easily) pollutants in tap water: chlorine, volatile organic chemicals (VOCs), and radon gas. Hot water creates more evaporation of these pollutants, especially when the water flows in a spray. Once these pollutants evaporate, you can be exposed to them by inhalation in addition to skin contact.

Frequent skin contact with chlorine can cause roughness, excessive dryness, and sometimes rashes, but few studies have been done on the long-term health effects of inhaling chlorine. However, it is estimated that more chlorine is absorbed through the skin during showers and bathing than by drinking chlorinated water. There's some evidence that drinking chlorinated water may have adverse impacts on health, so it's probably a good idea to reduce your exposure to chlorine from skin contact and inhalation.

However, the main health concern with volatile pollutants is inhaling VOCs or radon gas. To reduce chlorine, VOCs, or radon in bath water would require a costly whole-house water treatment system (described in chapter 11). Fortunately, there are many inexpensive shower filters that can reduce chlorine, VOCs, and radon. Almost all of the shower filters on the market contain a small canister of either granular carbon or redox media (often referred to by the brand name KDF), but rarely both. Carbon removes VOCs, radon, and most but not all chlorine that may be in shower water. Redox filtration removes all chlorine and reduces any pathogen growth in the filter. A few shower filters contain *both* carbon and redox media, providing the greatest reduction of all harmful volatiles. Three of these are recommended here.

Typical Shower Filter

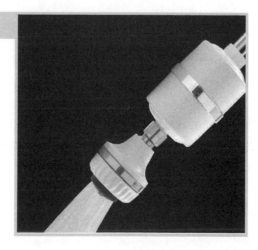

AQUASANA SHOWER FILTER with carbon and redox filters.

> *Price range:* $62–$70

> *Cartridge replacement:* $40, with estimated life of 6 months

PARAGON P2201 with carbon and redox filters.

> *Price range:* $40–$45

> *Cartridge replacement:* $20, with estimated life of 6 months

CRYSTAL QUEST SHOWER FILTER with carbon and redox filters.

> *Price range:* $35–$45

> *Cartridge replacement:* $17–$21, with estimated life of
> 6 months

Refrigerator Ice Maker Connections

Automatic RO systems, automatic distillers, and many filter systems can usually be connected to a refrigerator ice maker with just a few small parts that consist of tubing and connectors. Ice maker connection kits are inexpensive and typically cost about $10. Many hardware stores carry them. If you're a do-it-yourselfer and need a bit of help, search "how to hook up an icemaker" on www.ehow.com, and you'll find a helpful article.

Travel Filters for Treated Water

If you don't want to drink tap water while away from home, several small, portable water filters are available. These are for use with treated, uncontaminated tap water at motels, hotels, campgrounds, public faucets, and so on, not with untreated water from lakes, streams, and ponds. Two recommended models are listed here.

POWER SPORTS BOTTLE WATER FILTER. An inexpensive, lightweight 22-ounce polycarbonate bottle about 10 inches high by 3½ inches around. Block carbon filter removes chlorine, organic chemicals, lead and cysts and reduces bacteria.

Filter element NSF 53 certified.

Price range: $36–$42

Replacement cartridge: $24–$30, lasts for approximately 160 fillings

CWR TRAVEL FILTER GS1. A rugged, compact, stainless steel, crush-resistant unit about 7 inches high by 3 inches wide, with a 24-ounce capacity. Dual filters of granular carbon and ceramic. Removes chlorine, organic chemicals, lead, cysts, and bacteria.

Price range: $120–$160.

Replacement dual cartridge: $40–$50, lasts for approximately 2,000 gallons on treated water.

Camping Filters

As of this writing, only one model of camping filter will remove bacteria, cysts, and viruses from water—the First Need Deluxe. Other high-quality camping filters will remove essentially all bacteria and cysts, but not all viruses. Viruses are not a common health problem in outdoor water. But if you're looking for total protection and using camping filters that doesn't remove viruses, you should first treat the water with iodine or chlorine tablets. When the treated water is passed through a camping filter that includes a carbon element, the iodine or chlorine will be trapped and the output water will be safe and free of bad tastes. Since camping filters are designed for use with untreated water, they can also be useful in emergency situations when safe drinking water isn't available.

FIRST NEED DELUXE. This unit is suitable for small groups. It employs an ultrafine carbon matrix filter element that can be backflushed for cleaning. As of this writing, this is the only chemical-free camping purifier to obtain EPA certification as

First Need Deluxe Camping Filter

a microbiological purifier, meaning it removes bacteria, cysts, and viruses.

Price range: $75–$100

Replacement cartridge: $40–$45, lasts for approximately 125 gallons

MSR WATERWORKS II. This manual pump camping/emergency filter employs a ceramic filter element with a carbon block core plus an ultrafilter membrane to remove bacteria, cysts, and some viruses, plus organic pollutants.

Approved and recommended by the American Red Cross.

Price range: $100–$140

Cartridge replacement: $35–$50 each for the ceramic/carbon element and ultrafilter element. The ceramic cartridge may be scrubbed up to forty times for long life.

FIRST NEED BASE CAMP. Suitable for large groups and expeditions, this is an ultrarugged, high-capacity purifier in a stainless steel housing with sturdy metal components employs an ultrafine carbon matrix filter element that requires no cleaning.

Price range: $480–$560

Replacement cartridge: About $100, lasts for approximately 1,000 gallons

CHAPTER 13

Emergency Disinfection of Drinking Water and Storing Water

Hopefully, you'll never experience a drinking water emergency, when there's no tap water available or it has became tainted and unusable, but its important to be prepared. Government recommendations vary on the amount of water you should store for emergency use, from a mere three-day supply to a two-week supply. These figures translate to about 2 to 10 gallons per person in the household.

An effective way to store water for emergency use is to sterilize the water in storage containers using an appropriate amount of chlorine or iodine. Always keep a supply of chlorine bleach on hand (without added scent or fresheners); 5 to 7 drops of bleach will disinfect 1 gallon of water. Iodine comes in liquid and tablet forms and can be bought in sporting goods stores and drugstores. Iodinated water is for emergency use only—it is not recommended for prolonged daily use.

When filling bottles, add the appropriate amount of bleach or iodine (the latter as specified on the container), then fill the bottle completely to the top with water so no air remains when the caps are put on. Carefully seal the caps by wrapping tape tightly around them. When the caps are sealed, turn the bottles on their sides for a while so that the disinfected water contacts the insides of the caps and disinfects them as well. If an inexpensive, pour-through carbon filter is kept on hand, you can use it to remove the chlorine or iodine from stored water as it is needed.

If for any reason you aren't able to disinfect the water you are storing, consider purchasing a small *disinfecting* water filter (see Camping Filters in chapter 12). These are not typical filters; they are especially made to kill or trap pathogens. If you are bottling water for later use with a disinfecting filter, I recommend that you fill the bottles to the top with chlorinated tap water (tap water from a utility company). Disinfect the bottle caps in boiling water for one to two minutes, then put them on and seal them with tape.

In an emergency situation, if you can boil water or run a distiller, you can use either of those methods to disinfect water. But if you don't have electricity (often the case in an emergency), you'd need a nonelectric distiller. Neither boiling nor distilling is suitable for disinfecting water for storage, though. Both methods eliminate the residual chlorine in utility-supplied tap water, so unless you store the water in sterilized bottles, it won't remain sterile when it contacts the inside of the bottles.

Stored water should always be kept out of direct sunlight and stored in as cool a place as possible. If you are storing commercially packaged bottled water, replace it once a year. If you are storing water you've bottled yourself, replace it every six months. Date each container you store so you'll know when to replace it.

If you are out of water, remember there is a reservoir of 30 to 60 gallons in your hot water heater; there is a drain valve at the bottom of all water heaters, but before you drain a water heater, make sure the power has been off for a few hours to avoid being scalded. There's also a reservoir of water in each toilet tank (not the toilet bowl!). Don't drink water from water beds, because it contains toxic chemicals as preservatives. You can, however, use it for washing.

There are also many commercial products available for emergency drinking water, including prepackaged, bulk water in sterilized, airtight containers (for example, Aqua Blox). Enter "emergency drinking water" in any Internet search engine for further information on this subject.

A Radon Supplement

Radon is proving to be one of the most potentially harmful water pollutants. (For more information on the health hazards it presents, see chapter 2.) A radon map of the continental United States appears in chapter 3, but radon varies from site to site; some places within regions with high concentrations are free of radon, and it may be present in places outside the high-risk regions.

The removal of radon from drinking water by various methods of water purification is described in previous chapters. As explained, however, consuming radon in drinking water is not the greatest threat; inhaling radon and absorbing it through the skin are more serious dangers.

Radon gets into buildings in two ways. It can seep up through the ground and enter the air in the basement or foundation and spread through the rest of the house. In this case, radon gas must be prevented from getting into the house, and any radon already present must be removed from the air in the house. The remedies for this have to do with modifications to the foundation, air circulation, and so on, and are beyond the scope of this book, but many informative articles have been written on this subject.

The second way radon gets into the air in a house is by entering the tap water and evaporating into the air inside the house. In this case, the radon needs to be removed from the water, preferably before the water enters the house.

A first step is to contact your county health department and find out what tests have been done locally for radon and if any has been detected. If there is any suspicion about the presence of radon in your area, you can purchase an inexpensive kit for testing the air and water in your house. Radon test kits can be ordered online; just enter "radon testing" in any search engine to find suppliers. Test for radon in the air first. If the results indicate a high concentration of radon in the air in your house, further testing of both the air and the water will be needed. If the air in your house is relatively free of radon (radon is present to some degree everywhere), you don't need to test your water because the radon in water contributes only a small percentage of the radon in the air.

Test results are measured in picocuries per liter (pCi/l). The Environmental Protection Agency (EPA) recommends corrective action if radon in the air is 4 pCi/l or higher, and if radon in water is 4,000 pCi/l or more. If the radon concentration in your water is high, you should remove it by carbon filtration or by aeration. The type of carbon filter needed to remove radon from all of the water coming into the house is very different from a drinking water filter. (See chapter 11 for information on whole-house systems.)

CAUTION: A POE carbon filter will adsorb radon gas and itself become radioactive. If radon gas is present in water in a high concentration, the POE filter should be installed outside the house. If it must be positioned indoors, special shielding is required.

A low-cost way to partially reduce radon from household water is to install small carbon filters on the showerheads. This won't eliminate radon that evaporates from baths, cooking, a dishwasher, or a washing machine, but at least it will eliminate it from one area within the house that generates the highest radon concentrations. (See chapter 12 for shower filters.)

Atmospheric aeration is expensive, but it is the most effective and safest way to remove radon from water (aeration is described in chapter 7). An aerator for removing radon can be installed in a basement, garage, or utility room. A vent for removing the liberated radon gas is critical. In places where pipes don't freeze, an aerator may be installed outside, with no venting needed. Aerators typically cost about $2,000 to $5,000, including installation.

A private water system with an existing outdoor holding tank already has a potential aerator. You can aerate the water quite effectively by simply installing a fine spray nozzle where the water flows into the tank.

There are no EPA regulations regarding radon in air or water. States have varying requirements. To obtain more information about radon in the area where you live, go to "www.epa.gov/iaq//whereyoulive. html" and click on the link to your state in the list below the map. This will direct you to your state's radon office. For radon information in your local area, contact your county health department. The EPA telephone hotline for radon is 800-426-4791. The National Safety Council runs another useful hotline, 800-767-7236.

APPENDIX B

Glossary

absolute filter rating. Refers to the smallest particle size that a filter will trap 100 percent. For example, a 5-micron absolute filter will trap *all* particles 5 microns and larger. Also see *nominal filter rating.*

activated carbon. Carbon that has been roasted under special conditions to make it microporous and able to trap a variety of chemicals.

adsorption. A process by which molecules adhere to other molecules. This is how carbon filters trap chemicals.

aeration. The process of exposing water to large amounts of oxygen in order to remove certain kinds of chemicals.

aggressive water. Water that, because of its lack of minerals, tends to absorb chemicals from materials it contacts.

air gap. A vertical space within plumbing systems to prevent backflow from a drain or water line.

aquifer. A naturally occurring underground reservoir.

backwashing. Reversing the flow of water through a filter in order to cleanse it of accumulated particulate matter.

CA membrane. Cellulose acetate membrane; a type of membrane used in reverse osmosis that is not as effective as other types of membrane but which can tolerate chlorine.

carcinogen. A substance that causes or contributes to the onset of cancer.

chloramine. A compound used to disinfect water that is formed by a chemical reaction between chlorine and ammonia.

chlorine. A chemical used to disinfect water.

coliform bacteria. Harmless bacteria used as a marker to indicate the possible presence of pathogens.

colloid. A tiny particle that remains in suspension in a liquid.

condensate. Water that has been vaporized, then returned to liquid form. Same as *distillate*.

contact time. The amount of time that water contacts a filter medium while flowing through a filter.

contaminant. A substance in water that is harmful or otherwise undesirable. Same as *pollutant.*

corrosion. A chemical process by which water attacks metal surfaces and weakens or destroys them.

cryptosporidium. A microorganism that occurs in water in the form of cysts; a cause of gastrointestinal disorders.

deionization (DI). A process that removes minerals from water by ion exchange.

disinfection. The process of killing pathogens in water.

dissolved solids. Particles that have dissolved in water and are in solution. Also see *total dissolved solids.*

distillate water. Water that has been vaporized by boiling, then returned to liquid form. Same as *condensate.*

distillation. The process of boiling water, capturing the steam, and cooling it so that it condenses and purified water is produced.

effluent. The water flow that exits from a device or system.

feed, feedwater. A solution that enters a device or system for a specific purpose, as in a chlorine feeder.

filtrate. In a treatment device, the water flow after it has passed through a filter or membrane.

finished water. Water that has been improved by a water treatment plant and is ready to be delivered to customers.

flocculant. A substance added to water to make particles clump together in order to achieve better filtration.

Giardia lamblia. A microorganism that occurs in water in the form of cysts; a cause of giardiasis, a gastrointestinal disorder.

gpm, gph, gpd. The rate of water flow in gallons per minute, per hour, or per day.

groundwater. Water whose source is underground.

hardness. The amount of calcium and magnesium in water, minerals which, when levels are high, cause water to clean inefficiently and cause scale buildup.

heavy metals. The toxic metals in water, such as cadmium, lead, and mercury.

hydrogen sulfide. A toxic gas in water that smells like rotten eggs.

influent. The water flow input to a device or a system.

ion. An electrically charged atom.

ion exchange. A process by which undesirable ions in water are exchanged for more beneficial ones.

leaching. The process by which chemicals in the surface of a material enter the water that contacts that surface.

MCL. Maximum contaminant level; the maximum level permitted by federal law for a particular water pollutant.

MCLG. Maximum contaminant level goal for each MCL established by the EPA. The intent of MCLGs is to eventually and voluntarily lower the levels of water pollutants to those at which or below there is no known or expected health risk. MCLGs are currently nonenforceable.

medium/media. The material within a filter that enables it to improve water quality.

membrane. A thin film material porous enough to reject pollutants while allowing pure water to pass through.

MFL. Millions of asbestos fibers per liter of water; a measure of asbestos.

mg/l. Milligrams per liter; a measure of the amount of a substance in water. The equivalent of parts per million.

micron. A shortened term for one micrometer; one millionth of a meter.

mutagen. A substance that causes or contributes to genetic mutation.

NDWR. National Drinking Water Regulations, as established by the Safe Drinking Water Act of 1974.

NOC. Natural organic chemical.

nominal filter rating. Refers to the smallest particle size that a filter will trap most of. For example, a 5-micron nominal filter might trap 95 percent of all particles 5 microns or larger. Also see *absolute filter rating*.

oxidizing filter. A filter that uses a chemical reaction to remove pollutants.

ozone. A chemically aggressive form of oxygen used to disinfect water.

particulate. Particles in water.

pathogen. A microorganism that causes disease in humans.

permeate. That portion of water that passes through a reverse osmosis or other membrane.

pH. A measure of alkalinity or acidity in water. A pH between 1 and 7 is acidic, with the lower number being the most acidic; between 7 and 14 is alkaline, with the higher number being the most alkaline; and exactly 7 is neutral.

POE. Short for point-of-entry, referring to a whole-house water treatment system.

pollutant. A substance in water that is harmful or otherwise undesirable. Same as *contaminant*.

pore size. Refers to the smallest substance that a membrane will reject. For example, a 0.001-micron membrane might reject 99.9 percent of all substances 0.001 microns and larger.

potable water. Water fit for human consumption.

POU. Short for point-of-use, referring to a water treatment system in a specific location (such as a kitchen).

ppb. Parts per billion; a measure of the amount of a substance in water. One part per billion is equivalent to one-billionth of a gram per liter, or one nanogram per liter (ng/l).

ppm. Parts per million; a measure of the amount of a substance in water. Equivalent to milligrams per liter (mg/l).

Primary Drinking Water Standards. National drinking water regulations that pertain to harmful water pollutants.

psi. Pounds per square inch of water pressure.

raw water. Water before being subject to any treatment.

recovery rate. The ratio of pure water produced to total water used in the process of reverse osmosis.

redox. Short for reduction/oxidation, which is an exchange of undesirable for desirable ions in water.

regeneration. A process by which a filter medium is washed by reversing the flow of water through it. Same as backwashing.

rejection. The process whereby certain substances cannot pass through a membrane.

residual chlorine. The amount of chlorine that remains in water when it is delivered to its destination.

resin. A specially prepared mineral that is used in deionization and other kinds of water treatment.

reverse osmosis (RO). A process by which contaminants are rejected by a membrane while pure water is allowed to pass through.

scale. A hard buildup of mineral deposits on surfaces that contact water.

Secondary Drinking Water Standards. National drinking water regulations that pertain to the aesthetic and convenience qualities of water.

sedimentation. The settling of particulate matter in water.

semipermeable membrane. A thin film that allows passage of some materials while rejecting others.

SOC. Synthetic organic chemical.

softness. The quality of water that enables efficient cleaning and minimal corrosion, and which results from low amounts of calcium and magnesium.

teratogen. A substance that causes or contributes to birth defects.

TDS. Total dissolved solids; the standard measure of minerals dissolved in water.

TFC membrane. Thin film composite membrane; a type of high-performance membrane used in reverse osmosis and which cannot tolerate chlorine.

THM. Trihalomethane; a category of carcinogenic chemicals formed when organic chemicals in water combine with chlorine.

turbidity. A measure of the opacity, or cloudiness, of water.

ultraviolet disinfection (UV). A process by which intense ultraviolet light is used to kill bacteria and viruses in water.

VOC. Volatile organic chemical; a class of chemicals that evaporate easily. They can be absorbed through the skin or inhalation.

APPENDIX C

Resources

As of this writing there are about 31 million websites that have something to do with drinking water information. I've listed just a few of the useful ones below.

I have deliberately avoided listing websites that compare water purification methods (distillers, RO, filters, and so on) because I have found the majority to be unfair and to contain misinformation. For the same reasons, I have not listed websites that evaluate specific purifiers.

◆ General Information on Drinking Water

www.water-research.net/helpguide.htm
(provides links to many helpful websites)

◆ Bottled Water Information

www.bottledwater.org/public/faqs.htm
(pro bottled water)

www.nrdc.org/water/drinking/bw/chap4.asp
(anti bottled water)

www.dwrf.info/nrdc_bottled_water.htm
(neutral on bottled water)

www.bottledwater.org/public/contact
(for a list of IBWA-certified brands, click on "Brand List")

◆ Certification of Water Treatment Products

www.nsf.org/certified/dwtu/#complete

www.wqa.org

www.ul.com/water/prodcert/certification.html

◆ Water Testing Laboratories

www.ntllabs.com
(Watercheck with Pesticides Test)

www.watersafetestkits.com/html/drinkingkits.asp
(self-test kits)

www.puritec.com/store/category.cfm?Category-57
(self-test kits)

www.epa.gov/safewater/labs/ (follow links to state-by-state
listing of certified lab referrals)

◆ EPA Drinking Water Standards

www.epa.gov/safewater/mcl.html#mcls

◆ Information on Local Water Systems

www.epa.gov/enviro/html/sdwis/sdwis_query.html
(clicking on your state takes you to several options for finding
local information)

◆ Radon

To obtain more information about radon in the area where you live,
go to "www.epa.gov/iaq//whereyoulive.html" and click on the link
to your state in the list below the map. This will direct you to your
state's radon office. For radon information in your specific area,
contact your county health department.

◆ Unproven or Untested Water Concepts

www.aquatechnology.net/VITALIZER_PLUS.html (pro)

www.adermark.com/html/adr-4.html (pro)

www.chem1.com/CQ/clusqk.html (con)

INDEX

More Health and Nutrition Books
from Celestial Arts and Ten Speed Press

THE NEW DETOX DIET

The Complete Guide for Lifelong Vitality
with Recipes, Menus, and Detox Plans
by Elson Haas, MD
and Daniella Chace, MS, CN

"Represents the next wave in health
and healing—a must for everybody.
Detoxification is the missing link for
overall well being."
—Ann Louise Gittleman,
author of *Before the Change*

7 3/8 x 9 1/4 inches • 264 pages
$16.95 • 978-1-58761-184-1

THE TRANS FAT SOLUTION

Cooking and Shopping to Eliminate the
Deadliest Fat from Your Diet
by Kim Severson

"Belongs on every kitchen shelf."
—Marion Cunningham, author of
The Fannie Farmer Cookbook

6 x 9 inches • 144 pages
$12.95 • 978-1-58008-543-4

THE NEW OPTIMUM
NUTRITION BIBLE

Revised and Updated
by Patrick Holford

"Optimum nutrition is the medicine
of the future."
—Linus Pauling, two-time
Nobel Prize winner

6 x 9 inches • 576 pages
$19.95 • 978-1-58091-167-2

THE FASTING HANDBOOK

Dining from an Empty Bowl
by Jeremy Safron

Raw food pioneer presents a variety of
detoxifying and healing techniques to
enhance the cleansing process.

6 x 8 inches • 128 pages
$11.95 • 978-1-58761-231-2

Available from your local bookseller, or order direct from the publisher:

TEN SPEED PRESS
www.tenspeed.com
order@tenspeed.com
800-841-BOOK